On the Arctic Frontier

Ernest deKoven Leffingwell in 1914.
Deborah Storre Collection

On the Arctic Frontier

Ernest Leffingwell's Polar Explorations and Legacy

JANET R. COLLINS

Washington State University
Pullman, Washington

Washington State University Press
PO Box 645910
Pullman, Washington 99164-5910
Phone: 800-354-7360
Fax: 509-335-8568
Email: wsupress@wsu.edu
Website: wsupress.wsu.edu

First printing 2017

Library of Congress Cataloging-in-Publication Data

Names: Collins, Janet R. (Janet Ruth), 1952- author.
Title: On the arctic frontier : Ernest Leffingwell's polar explorations and legacy / Janet R. Collins.
Description: Pullman, Washington : Washington State University Press, [2017] | Includes bibliographical references and index.
Identifiers: LCCN 2017022065 | ISBN 9780874223514 (alk. paper)
Subjects: LCSH: Leffingwell, Ernest de K. (Ernest de Koven), 1875-1971. | Explorers--United States--Biography. | Polar regions--Discovery and exploration.
Classification: LCC G585.A6 C65 2017 | DDC 917.98/7043092 [B] --dc23
LC record available at https://lccn.loc.gov/2017022065

On the cover: America, Library of Congress Prints and Photographs Division; sled dogs, USGS Photographic Library, Leffingwell Collection; Ernest Leffingwell, Deborah Storre Collection

In loving memory of my mother,

Ruth C. Collins

who loved to travel, had a wonderful sense of humor,
a spirit for adventure, and a sincere appreciation for other
cultures and landscapes.

Contents

Maps

Preface

This project started with an adventure and a name on a map. Backpacking with friends in the Sadlerochit Mountains of the Arctic Refuge in July 1992, I noticed "Camp 263" on my map. As a geographer and map librarian, the name sparked my curiosity. When I returned home I consulted *Orth's Dictionary of Alaska Place Names* for "Camp 263" and learned of Ernest Leffingwell.

Leffingwell and the place that so defined his life's work fascinate me. That place was later defined as the Arctic National Wildlife Refuge (Arctic Refuge). There has been very little written about Leffingwell, or his prominent role in exploring that region, and he deserves recognition for his accomplishments. Leffingwell defined and mapped the coastline of northeastern Alaska, mapped the geography and geology of a significant part of the Arctic Refuge, reported the oil seepage at Cape Simpson in U.S. Geological Survey Professional Paper 109, and named the formation that underlies the Prudhoe Bay Oil Field. He is also credited with pioneering work on ground ice, also known as permafrost.

My fascination with the Arctic began in the early 1970s while enrolled in a college course titled "Arctic Environment." My July 1992 backpacking adventure led to research on Camp 263 and Leffingwell: When was he there? Why was he there? What kind of logistics were involved in getting there? And what kind of experiences did he have?

Orth's *Alaska Place Names* credited the name to Leffingwell and Professional Paper 109, titled *The Canning River Region Northern Alaska,* published in 1919. I examined the publication and found it interesting, but other commitments precluded further investigation. Over ten years later I decided to present a paper about Leffingwell at a conference of map librarians in Alaska. In preparation I visited Dartmouth College where the Leffingwell papers reside, and the U.S. Geological Survey office in Anchorage, Alaska, and was amazed by Leffingwell's accomplishments.

When I retired in 2008, I finally had the time to revisit the Arctic Refuge and dig into the research. I returned to Dartmouth in the fall of 2008, and over the next four years I retraced Leffingwell from his birthplace in Knoxville, Illinois, to the University of Chicago where he worked toward a PhD in geology. Research continued throughout the United States and Europe.

Leffingwell was part of the Baldwin-Ziegler Expedition, an attempt to reach the North Pole in 1901–1902. While researching the expedition, I traveled to the Library of Congress to examine the Baldwin Papers and the Barnard Papers. Eventually, I visited the Smithsonian Archives, The Explorers Club, American Geographical Society, Royal Geographical Society in London, Danish Arctic Institute, and Royal National Library of Denmark in Copenhagen. In addition, I visited the University of Victoria and the Victoria City Archives in British Columbia, Canada, and the National Archives of Canada in Ottawa. Contact with Leffingwell descendants proved invaluable in learning more about him.

Research proved Leffingwell a far more interesting and complex subject than I had imagined. He led an unconventional life in many ways, yet his upbringing was one of privilege and far from unconventional. Leffingwell was always interested in learning, but once his work in the far north was completed in 1914, he never returned to the Arctic or settled into a career. He was driven to make contributions to scientific knowledge and had produced high quality work in Alaska that remains relevant today. His contributions to scientific knowledge ended with his work in Alaska and the previously mentioned Professional Paper 109. His long and full life ended at age 96 in late January 1971.

Introduction

This biography, set primarily in the northeastern corner of the State of Alaska and within the boundaries of the Arctic National Wildlife Refuge, tells of a man and a place during the years 1906 to 1914. Ernest deKoven Leffingwell made three trips to northeastern Alaska, spending nine summers and six winters conducting field research and mapping. Leffingwell was primarily a geologist, but also a geographer, cartographer, and naturalist who enjoyed reading and learning languages. He took measure of the land, but the land also took measure of him. The environment tested his determination, flexibility, and resiliency, his problem-solving abilities, and his dedication and commitment to the pursuit of scientific knowledge.

Leffingwell estimated that he traveled 4,500 miles by sledge or boat on 31 trips during his field work, which extended over 30 months. He pitched camp about 380 times, living mainly in a tent. It was an amazing feat.

Leffingwell made significant contributions to exploration and scientific studies, not only in Alaska, but as the chief scientific officer of the earlier Baldwin-Ziegler expedition to Franz Josef Land. His experiences, his work on ground ice (permafrost), and his defining, naming, and mapping of the geography and geology are the subject of this work, as well as an explanation of the relevance of his work today. His work has been ignored for much too long.

The logistics, and his trials and tribulations in traveling to and from northeastern Alaska, were formidable, and recounting them adds to our understanding of the big picture of Arctic exploration. Tremendous time and energy were required to maintain shelters and clothing, to obtain food to supplement what had been carried north, and to survive in adverse weather conditions. Although Leffingwell accomplished much of the work on his own, this book also recognizes the contributions of the Inupiat. Leffingwell's work could not have been completed without their assistance.

Working conditions were particularly challenging in winter because of wind, cold, and the lack of light. The sun set in November and did not rise again until late January. The winter's cold and relentless wind prevented Leffingwell from using his instruments, which functioned best in conditions of little or no wind, as accuracy was susceptible to the slightest movement. In the summer, the sun circled overhead for 24 hours a day. Spring and summer brought shifting pack ice, melting snow and ice, and hungry mosquitoes. And there was fog and rain, and more wind. In spite of many challenges, he managed to accomplish an impressive amount of work in brief windows of opportunity.

Through Leffingwell's accounts, and accounts of him by others, we gain insight into the qualities that defined him. These sources help reveal his values, morality, and his interactions with others, including the Inupiat, whalers, traders, fellow scientists, and other explorers. Through them we gain a sense of his daily life while on the North Slope, much of it conducted entirely at his own expense. Those experiences impacted his life forever.

The primary collection of Leffingwell source material is housed at Dartmouth's Rauner Library as part of the Stefansson Collection, including his eight journals covering the Baldwin-Ziegler Expedition and his work in northeastern Alaska. Additional journals of the Baldwin-Ziegler Expedition were accessed at the Library of Congress, including those of Evelyn Briggs Baldwin and Leon Barnard.

Leffingwell's participation in the Baldwin-Ziegler Expedition, recounted in the first chapter, was his introduction to the Arctic and provided the foundation for his later work in Alaska. The expedition taught Leffingwell about the Arctic environment, survival skills, operating scientific equipment, competing agendas, expedition organization, and challenges of a wide variety. His surveying and mapping was critical to Baldwin's efforts to map Franz Josef Land. A friendship with one of his shipmates, Ejnar [Einar] Mikkelsen, a Dane, resulted in the Anglo-American Polar Expedition (AAPE), co-commanded by Leffingwell and Mikkelsen—their first trip to Alaska. Their pack ice trip off of northern Alaska in 1907 represented the most hazardous event during Leffingwell's time in Alaska.

The Leffingwell story relies on his own journals and on published and unpublished writings by Alice Barnard, Vilhjalmur Stefansson, and Ejnar Mikkelsen, Leffingwell family papers, and memories of two of Leffingwell's grandchildren. Selected journals by members of the

Canadian Arctic Expedition at the National Archives of Canada were also used as sources.

Many of Leffingwell's journal entries are in pencil, and he frequently used abbreviations and initials. He was meticulous in detail, including specific times of arrivals and departures from various locations. They have been generalized in this work. Quotes from his journals are included to give the reader a sense of his challenges, frustrations, and triumphs.

His journals elaborated on weather conditions and temperatures. Leffingwell used both Fahrenheit and Celsius readings, often without telling us which he was referencing. (See the conversion chart in Appendix 5, or visit www.srh.noaa.gov/ama/?n=conversions#tables.) He also carried separate field notebooks that became the geologic basis for Professional Paper 109. Additional source material was found in personal correspondence and news articles.

Distilling Leffingwell's highly technical scientific report covering cartography, geography, and geology into language understandable to a reader unfamiliar with those disciplines was challenging. Because I am not a geologist, readers with more specialized knowledge are encouraged to delve into the primary literature, especially Professional Paper 109, available on the internet at www.dggs.alaska.gov/pubs/id/3803.

Historical Timeframe—Context and Perspective

Leffingwell's exploration coincided with a time of great interest in the Arctic, Antarctica, and the Northwest Passage, and in reaching the North and South Poles. At the turn of the twentieth century explorers were busy making their marks in those regions, and numerous expeditions were launched with wide public support. Robert Peary reached the North Pole in 1909. Ernest Shackleton made his first trip to Antarctica in 1907 and came within 100 miles of the South Pole in 1909. Roald Amundsen completed the Northwest Passage in 1906 and then reached the South Pole in 1911.

Beyond these "firsts," the Arctic still held unexplored and uncharted areas. One of those was north of Alaska. Geographers, explorers, and cartographers hotly debated whether there was land directly north of Alaska. Rumors abounded of non-existent Arctic places referred to as "Keenan Land" and "Crocker Land." Leffingwell and Mikkelsen chose

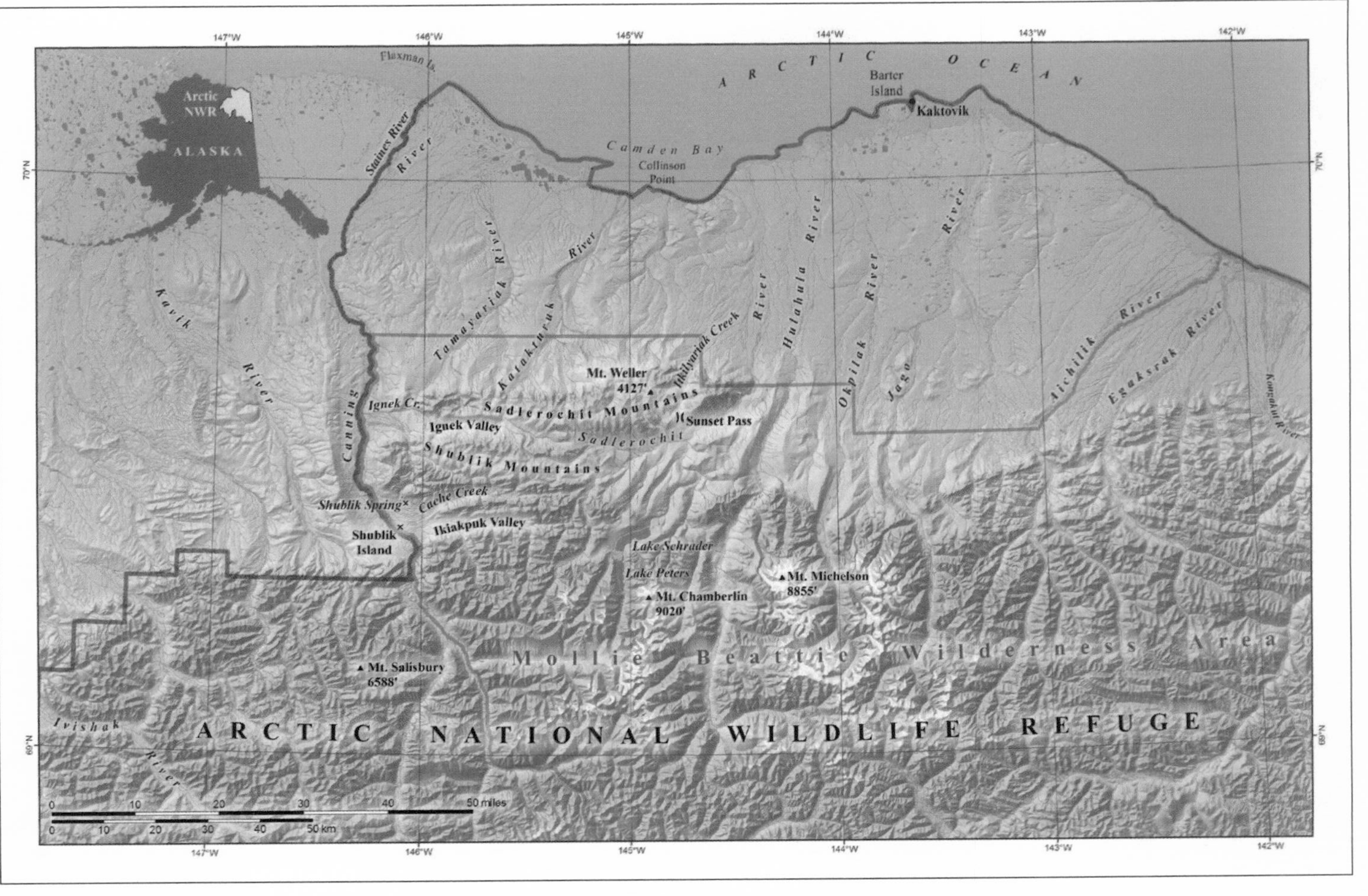

Northwestern Arctic National Wildlife Refuge. *U.S. Fish & Wildlife Service, Region 7, Division of Realty, Scott McGee, cartographer.*

to explore the area north of Alaska in hopes they could help resolve the controversy.

Another dynamic of the era was a movement to add to scientific knowledge that started with the First International Polar Year in 1881–1882. It was a movement about which Leffingwell felt strongly. The United States government also wanted to map potential mineral and natural resources for future development throughout the country and in Alaska.

It appeared a small world for explorers, even at the turn of the twentieth century. The Royal Geographical Society, American Geographical Society, and The Explorers Club brought explorers together. The societies themselves may not have had significant financial resources, but their wealthy supporters and prominent members did. That distinguished group included Walter Rothschild, Alexander Graham Bell, and the Duchess of Bedford. Financial support by the societies and their members was critical to the launching of more than one expedition, including the 1906 Anglo-American Polar Expedition organized by Mikkelsen and Leffingwell, to which all of them contributed.

The region of Leffingwell's exploration in northeastern Alaska had been home to the Inupiat for millennia. Only a few outsiders had explored it, notably Sir John Franklin as part of his Second Overland Expedition. Franklin traveled the coastline in 1826, mapping and naming a few features (including Flaxman Island) as he traveled west toward Barrow before returning east to the Mackenzie River. Certainly whalers were familiar with the area, but they did not map or explore inland.

Leffingwell was thus an early contributor to inland and coastal exploration and knowledge. Yet the efforts of Mikkelsen and Leffingwell were overshadowed by more prominent expeditions, including those of Amundsen, Peary, Shackleton, and Stefansson, which in large part explains why Leffingwell and his work are not well-known.

Arctic National Wildlife Refuge

Understanding Leffingwell and his Alaskan experiences demands an understanding of the environment in which he worked. The setting for his explorations between 1906 and 1914 is what is now called the Arctic National Wildlife Refuge, commonly referred to as the Arctic Refuge or ANWR. Throughout the United States, the Arctic National

Wildlife Refuge is perhaps best known for its wilderness and potential for petroleum resource development. It is bounded on the north by the Beaufort Sea, on the east by the Yukon Territory of Canada, on the west by the Canning River, and on the south by Arctic Village and its adjacent areas north of the Yukon River.

One of the largest and most important wilderness areas within the United States, the Arctic Refuge is regarded by many as America's Serengeti, with an intact ecosystem and plentiful, diverse wildlife. Preservation of open spaces like the Arctic Refuge evokes tremendous passion between those who value its ecological and environmental significance and those who see its potential for oil and gas development. Stakeholders include the Inupiat, the indigenous people of the North Slope, government agencies, scientists, academics, recreational users, environmentalists, businessmen, and politicians. To many it embodies wilderness values, so well defined and expressed by Roger Kaye in his book *Last Great Wilderness*. For others it presents economic opportunities that outweigh preservation concerns. These conflicting interests make the Arctic Refuge a prominent political issue.

The diversity of wildlife within the Arctic Refuge includes grizzly bears, polar bears, caribou, Dall sheep, fox, moose, musk oxen, and wolf. During the intense summer season the sun circles overhead and bathes the landscape with 24-hour sunlight. Plant life explodes in carpets of many colors. Birds migrate from Antarctica, South America, India, and China to the refuge each year to nest, with 201 species recorded.

During winter the sun does not rise and temperatures in the coldest months average -25° Fahrenheit at Kaktovik to -33° Fahrenheit at Arctic Village. Wind chill significantly lowers those temperatures. Polar bears disappear to their dens while the few remaining musk ox traverse the coastal plains seeking nourishment.

The Brooks Range, an extension of the Rocky Mountains, transects the Arctic Refuge. The U.S. Geological Survey shows the highest peak, Mount Chamberlin, at almost 9,000 feet.[1] The highest mountains are glaciated and major drainages trend north-south from the Brooks Range. In the area Leffingwell mapped, the geology is a mix of granitic and metamorphic rocks in the Brooks Range, and limestone, shale, and sandstone north of the range. Traveling north of the Brooks Range toward the Beaufort Sea, the mountains give way to foothills, undulating hills, then the relatively flat coastal plain. Precipitation is about six to eight inches per year, and with few exceptions trees are non-existent. This contrasts to the area south of the Brooks Range where precipitation is over nine inches per year, vegetation can be dense, and spruce trees occur throughout the landscape.

The Inupiat have lived and hunted north of the Brooks Range in and near the Arctic Refuge and utilized its resources for many generations. They continue to do so to this day. The Gwich'in to the south have done the same. It is important to recognize that indigenous peoples knew and explored the area long before white men arrived. Today, the two major communities are Kaktovik on Barter Island in the Beaufort Sea and Arctic Village south of the Brooks Range. Recent census figures list their populations at 239 and 152 respectively.

We have followed here the guidelines of the Alaska Native Language Center at the University of Alaska Fairbanks in the use of Inupiat and Inupiaq (www.uaf.edu/anlc/languages/i). Briefly, Inupiaq—meaning "real or genuine person"—can refer to a person of this group ("he is an Inupiaq"), and can also be used as an adjective ("she is an Inupiaq woman"). The plural form of the noun is Inupiat, referring to the people collectively.

Leffingwell's work extended over the northwestern quadrant of the Arctic Refuge. His work spanned the area from the Beaufort Sea to south of the Brooks Range at latitude 69°, bounded on the east by the Yukon border, and on the west by the Canning River. Leffingwell spent part of the nine summers and six winters based at Flaxman Island in the Beaufort Sea, while he explored and mapped much of the area within and north of the Brooks Range that now lies within the Arctic Refuge.[2]

In 1960, forty-six years after Leffingwell departed the North Slope for the last time, the Arctic National Wildlife Range was created, setting aside 8.9 million acres "For the purpose of preserving unique wildlife, wilderness, and recreational values."[3] It was renamed the Arctic National Wildlife Refuge in 1980. Today it encompasses slightly fewer than 20 million acres, roughly the size of South Carolina.

Between 1906 and 1914 Ernest deKoven Leffingwell explored this region, initiating an interaction between man, science, and the environment that deepens our understanding of northeastern Alaska.

Prologue

On a chilly November evening in 1897, Norwegian Fridtjof Nansen, perhaps the most famous explorer of that time, spoke at the University of Chicago about his recent experiences in the Arctic. A 22-year-old geology graduate student by the name of Ernest deKoven Leffingwell sat in the audience.

Nansen engagingly spoke of the adventure and the challenges of Arctic travel, as well as the physical beauty of the region and the importance of scientific work. His words and presentation had a profound effect on Leffingwell, and provided the sense of direction he had been seeking. He had changed his mind a number of times about what he wanted to do with his life. Early on he had considered becoming a clergyman like his father, then a physician, biologist, or physicist, before settling on the discipline of geology. Leffingwell was captivated by what he heard and he later remarked that his passion for the Arctic began that evening.

Nansen's lecture related his 1893 to 1896 experiences while intentionally adrift in the pack ice of the Arctic Ocean, eventually emerging in the North Atlantic to study ocean currents. He was commander of the expedition conducted aboard the *Fram*, a ship he had specially designed for the work. During March 1895 Nansen left the ship with a companion in an attempt to be the first to reach the North Pole. After traveling north for 140 miles, he gave up the quest, returned south, and overwintered in Franz Josef Land north of Russia.

Nansen's mission was to explore unknown regions and gather scientific information. He believed that, no matter which country attempted to reach the North Pole, such efforts should be encouraged by all. Nansen strongly encouraged scientists to come forward and contribute to the body of knowledge about the Arctic. A visionary who maintained that in order for science to grow, it needed to solve problems, Nansen felt that those problems were waiting in the Arctic. His words, including a quote from Alfred Lord Tennyson's poem "Ulysses," were eloquent and enticing:

> The history of humanity has been a continuous struggle from darkness towards light, but one of the proudest chapters of the history of that

> struggle for light is perhaps the chapter about polar exploration. You will nowhere find finer pioneers for science described—men in whom you find the finest, noblest features of human character; men with 'one equal temper [of] heroic hearts, strong in will to strive, to seek, to find, and not to yield.'[1]

Four years later Ernest deKoven Leffingwell answered that call by participating in the Baldwin-Ziegler Expedition of 1901–1902 as chief scientific officer. It would prove a defining moment in his life.

CHAPTER 1

To the Arctic: Leffingwell and the Baldwin-Ziegler Expedition

Ernest DeKoven Leffingwell's passion for the Arctic was sparked at age 22 when he attended Fridtjof Nansen's University of Chicago lecture. Hearing the famed Norwegian explorer describe a way of life that combined his love of science and the outdoors was a defining moment for Leffingwell. His first opportunity to travel north would come four years later with the Baldwin-Ziegler expedition, the goal of which was to be the first to reach the North Pole and contribute to the scientific knowledge of the Arctic. The expedition was to depart in 1901 under the command of Evelyn Briggs Baldwin, a meteorologist and an aspiring polar explorer. William Ziegler, a New York multi-millionaire, was its financier.

Baldwin's Plan

Born in 1862, Baldwin spent most of his youth on a farm in Kansas, and early on developed a love of geography and desire to explore. He traveled in Europe, then taught and worked as a school administrator in Oswego, Kansas. In the early 1890s Baldwin became an assistant observer for the U.S. Weather Bureau, then served as a meteorologist with Robert Peary's second expedition to Greenland in 1893. In 1898 he was meteorologist and second in command under Walter Wellman in an expedition that explored the archipelago of Franz Josef Land. Franz Josef Land had been partially mapped, and Baldwin was interested in completing the job. In 1899 he began lecturing and raising funds to organize and lead his own expedition.

Baldwin's financial support came from William Ziegler, the head of a baking powder trust, which included the Royal Baking Powder Company. Their contract, drawn up in October 1900, established their goal of "the attainment of the North Pole and the exploration of any land that may be found in its immediate vicinity."[1] For Ziegler, it was only about obtaining the North Pole for the United States, and at any cost. Baldwin decided to make the attempt for the pole via Franz Josef Land. With the exceptions of Greenland and Ellesmere Island, Franz Josef Land offered the highest latitude and was regarded as a more desirable route for North Pole attempts. Named by the Russians as *Zemlya Frantsa Iosifa*, Franz Josef Land is administered by Russia, and consists of 191 islands northeast of Svalbard and about 800 miles north of Murmansk, Russia, between latitudes 80° and 82° north. Discovered in 1873 by Karl Weyprecht and Julius von Payer, the archipelago was explored by Leigh Smith in 1880 and 1881, Frederick Jackson in 1894, Fridtjof Nansen in 1895 and '96, and Wellman in 1898. Several of those expeditions were attempts for the pole. The Duke of Abruzzi reached the northernmost island, Crown Prince Rudolf Island, in 1899.

Baldwin carefully outlined his plan for the expedition but shared it only with his closest friends. The secrecy carried over in the expedition, where he kept his plans to himself until the last possible moment.

Once the expedition reached the highest possible latitude in Franz Josef Land, Baldwin planned to establish depots of provisions (caches) and travel to the pole via dog sledges constructed aboard the ship. Fifteen to twenty men would comprise the party, in a trip that would require seventy-five to one hundred days. A support team would accompany the polar party to between latitudes 85° and 86° north and then return to camp. Baldwin calculated (based on Peary's expedition to Greenland) that they could travel a minimum of 34 miles per day. Assuming this could be maintained, it would take fewer than 60 days to travel to the pole and return to Franz Josef Land, an overly optimistic determination even in the best of conditions.

The total estimated cost was $70,000 (close to two million in 2016 dollars when adjusted for inflation.). Baldwin later indicated that $142,000 of Ziegler's money was spent on the expedition.[2] An article in the December 28, 1903 edition of the *Brooklyn Daily Eagle* (partly owned by Ziegler), reported "nearly half a million dollars" spent.[3] Whichever figures were correct, it was an enormous amount of money. (To estimate costs in 2016 dollars adjusted for inflation, multiply 1910 costs by 25.)

Baldwin acquired a former Scottish whaling vessel that he named *America*, and two supply ships, the *Frithjof* and the *Belgica*. Used by the Wellman expedition in 1898, the *Frithjof* would accompany the *America* to Franz Josef Land. The *Belgica* would travel to Shannon Island off the eastern coast of Greenland and deposit a cache before returning to Europe. That location was chosen based on Baldwin's understanding and projection of ice flow movement. Once the North Pole had been reached, the ice floes would likely carry them toward Greenland, where they could resupply before returning to Franz Josef Land.

Leffingwell Applies to the Expedition

Leffingwell learned of the Baldwin-Ziegler expedition from press coverage and wrote to Baldwin. A graduate student in physics and geology at the University of Chicago, two of his professors, Albert Michelson and Samuel W. Stratton, wrote letters on his behalf. (Leffingwell would later name a mountain for Michelson, and Stratton became head of the National Bureau of Standards.)

Not until January 1901 did Baldwin consider the applications for staff, a late time frame considering the expedition would depart Scotland in late June of that year for Norway. Leaders of Arctic expeditions generally sought strong, athletic, adventurous, outdoor types when selecting their members. They needed to be tough enough to withstand the rigors of the Arctic, unknown challenges, inevitable hardships, and face the possibility of not returning. Leffingwell was well suited.

His application included a biographical sketch. He was born January 12, 1875, in Knoxville, Illinois, to Charles and Elizabeth (Francis) Leffingwell. His parents co-managed a school for girls, St. Mary's, in Knoxville, and later, a boys' school, St. Alban's. As a youth and member of a prominent, wealthy family, Leffingwell loved camping, fishing, boating, and going "barefoot all summer" in northern Michigan near Old Mission. He loved being outdoors, a lifelong passion. Leffingwell was a natural athlete who excelled at track while at Trinity College from 1894 to 1895, and while attending the University of Chicago. Once he joined the Illinois Naval Reserve track team, he quickly became the captain. Leffingwell was educated at Racine Grammar School (an Episcopal preparatory school also known as DeKoven's School for Boys) in Racine, Wisconsin. He continued his education at his father's school, St. Alban's.

He entered Trinity College in Hartford, Connecticut, in 1892, as a six-foot-tall, 175-pound youth with dark hair and piercing blue eyes.

Like his father, Leffingwell planned to pursue theology, gave that idea up, and briefly attempted medical college at Columbia University. His studies emphasized physics and math. No transcripts exist of his studies at Trinity, but he graduated in 1895 with an undergraduate degree, and later, a master's degree.[4]

When he entered the University of Chicago in fall 1896 to pursue a PhD, he took courses in biology, then physics, before settling on geology. Like his father, Leffingwell's interests varied widely. He was socially active and a member of the American Academy for the Advancement of Science, Chicago Geographic Society, Washington Geologic Society, the Mandolin Club, and the Chicago Quadrangle Club. Leffingwell was also affiliated with the fraternity, Sigma Xi.

Before hiring Leffingwell as his chief scientific officer, Baldwin requested that Leffingwell specifically address his understanding of astronomy, skills in cartography, surveying, and use of instruments including the theodolite (used in surveying to measure horizontal and vertical angles), and the sextant (used to determine latitude). Baldwin was also interested in Leffingwell's background in the arts, including photography, music, and gymnastics.

Another consideration was how expedition members would adapt to Baldwin's command structure. Leffingwell, who had served in the United States Navy during the Spanish American War, understood chain of command. He served as a seaman on the battleship USS *Oregon* between June and September 1898 and was a gunner at the battle of Santiago.[5] After his return from active duty, Leffingwell resumed his studies at the University of Chicago until joining the Baldwin-Ziegler expedition.

Leffingwell was hired as expedition geodesist, responsible for taking precise location measurements of the earth's surface, but his title was chief scientific officer. Wages were $25 per month. In March 1901 he left the University of Chicago for training in the use of navigation instruments at the U.S. Coast and Geodetic Survey in Washington, DC. In addition he studied astronomy and surveying. Leffingwell remained in Washington until the end of May 1901. His contract with the Baldwin-Ziegler expedition was not signed until May 24, 1901, in Dundee, Scotland, though he had begun working for Baldwin in March. Concerned with other expedition details, Baldwin left the signing of contracts until just prior to sailing. His lack of attention to the men who were to accompany him offered fair warning about the future of the expedition.

While at the U.S. Coast and Geodetic Survey Leffingwell spent endless hours writing letters regarding the purchase or loan of equipment, reporting to Baldwin on his progress, organizing the shipping and packing, and developing a work plan for the scientific component of the expedition.

Leffingwell's writing was circumspect, especially to the companies providing the instruments. He highlighted the long-term financial benefits of testing and endorsing the performance of the company's equipment in the Arctic environment. A March 25, 1901, letter to Berger & Sons in Boston, about engineering and surveying instruments, offered: "One company is loaning $2500 of valuable instruments, simply because our records will come before the scientific world and be severely criticized. If the work done passes inspection the instrument maker will have a recommendation that no one can doubt."[6]

In providing the details for those instruments specifically constructed for the expedition, Leffingwell modified the original designs to accommodate the Arctic cold and wind. Once the instruments arrived, he spent hours calibrating and checking them for accuracy. Other instruments were purchased or loaned by U.S. government agencies including the Coast and Geodetic Survey, the Signal Corps, and the Navy Bureau of Equipment. The instruments included a surveying camera, transits, compasses, chronometers, a sounding apparatus, altazimuth, theodolite, magnetometer, thermometers, and hydrometers. A chronometer

Leffingwell in 1898 while in the U.S. Navy, serving aboard the USS *Oregon*. *Deborah Storre Collection*

is a precision instrument used to measure time—critical in determining longitude. An altazimuth is used to observe angles of celestial bodies. A magnetometer measures the strength and direction of the earth's magnetic field, while a hydrometer measures rainfall and other characteristics of water including density. In addition to the instruments, there were chemicals, batteries, wires, switches, tapes, and endless small items to be purchased and inventoried.

In addition to learning how to use the various instruments, Leffingwell spent many of his evenings reading and studying reports from many earlier Arctic expeditions including those of Fridtjof Nansen, Robert Peary, Walter Wellman, Adolphus Greely, and Frederick Jackson. He read about the Challenger expedition of 1872–1876, which studied oceanography. His studies included "moon culminations and occultations," a methodology used to determine longitude and geographic location. Occultations are a temporary celestial eclipse, which occurs when one object is visible but blocks out another that is located further away—for example, when the moon passes in front of a star.

Raised in an environment that emphasized education and hard work, it is not surprising that Leffingwell's work plan—which included astronomy, magnetism, meteorology, and hydrography—and list of instruments was ambitious. By the end of March, Leffingwell had met with zoologists at the Smithsonian Institution to discuss a variety of topics, including marine biology, ornithology, and mammalogy. They helped plan his upcoming work and provided appropriate literature. He wrote Baldwin about their positive response to the expedition, the acquisition of marine specimens, and the fame it might bring to the expedition. He was eager to prove himself.

Departure for Scotland

On Sunday, May 19, Leffingwell packed his trunks and took the train from Washington to New York. The following day he picked up equipment that had been shipped to New York, tuned the chronometer, and spent time at Ziegler's office. That evening, expedition members attended a farewell dinner hosted by the Arctic Club in their honor.[7]

Leffingwell shared his optimism about the expedition with his family. His father wrote to Baldwin on June 1, 1901, "I am much gratified to hear from my son that you have won his confidence and admiration."[8] Long before the expedition ended those sentiments would change dramatically.

By Saturday, May 25, Leffingwell was aboard the *Astoria* sailing for Glasgow, Scotland. He noted the cold and the rough seas. Leffingwell suffered severe seasickness and was unable to work for 11 days of the crossing. Seasickness plagued him throughout his life. The weather improved two days prior to their docking in Glasgow in the evening on Tuesday, June 4. He took a room at the Grand Hotel, grateful to be off the ship. Leffingwell then made his way to the expedition ship *America*, docked in Dundee, arriving June 6. One of the first people he met was Ejnar Mikkelsen-Loth.[9] Leffingwell was impressed by his experience on the Amdrup expedition and his skills as a cartographer.

Within a few days, Leffingwell wrote a lengthy letter to Baldwin in London about conditions aboard the *America*. He was forthright with his opinions and made suggestions to Baldwin about matters he felt needed attention including the food, lack of work performed by some, and the sleeping accommodations. Leffingwell had an air of self-confidence and, at age 26, a bit of youthful arrogance. His expectations of expedition conditions were high and perhaps unrealistic, and would resurface throughout the journey. He suggested that Baldwin arrive before sailing to talk with the ship's captain, officers, and crew about problems he felt should be resolved. He not only suggested a meeting but that it be a "kind talk" that would improve "humor."

While Leffingwell's intent was honorable, it is unlikely that others on the expedition were as outspoken. Baldwin did not make the trip and never dealt with the issues. Leffingwell clearly felt the issues were important and needed to be addressed. If he could improve conditions, solve problems, and make things easier, why shouldn't he? Throughout the expedition, he continued to express his opinions to Baldwin, which undoubtedly contributed to tension between them

Leffingwell turned his attention to the trip preparation. The days before sailing were long and busy. He studied in the evenings and recorded his progress in his journal. He continued refining the instruments, often with Mikkelsen's assistance. Both Leffingwell and Mikkelsen worked hard, and were fully committed to the expedition's success.

The group was warmly received in Scotland. The Royal Observatory in Edinburgh offered its assistance to Leffingwell setting up the ship's pendulum. It was critical to have accurate time in order to establish their geographic position as they traveled. Leffingwell also worked on a chronometer journal and helped Russell Porter, the artist, draw a circumpolar map. His evenings were often spent reading books, including Frederick A. Cook's *Through the First Antarctic Night*, about an 1898–1899 expedition below the Antarctic Circle, and another about Swedish explorer S. A. Andree's 1897 attempt to reach the North Pole in a balloon. He studied expedition accounts by Nansen and Frederick Jackson, and he wrote letters home.

Danish explorer Ejnar Mikkelsen, circa 1906. *Deborah Storre Collection*

Shortly before midnight on June 28, the expedition sailed from Dundee for Tromso, Norway. Just prior to sailing, Baldwin had drawn up an Addendum to expedition members' contracts stipulating that they were not allowed to bring or use any photographic equipment and furthermore, no member could share information about the expedition unless given permission by Baldwin or his designee. Baldwin sought to totally control the flow and content of information.

Crossing the Arctic Circle

Crossing the Arctic Circle at latitude 66°30'north was a cause for celebration with a traditional initiation process. The "initiation" took place off the coast of Norway and was performed for all those who had not previously "been across the circle."

Leffingwell's journal for July 3 noted: "Bright fine day. Many fin whales spouting near. (200m) Adjusted large sextant and gave it to Menander. a.m. Salt water bath a.m. Loafed p.m. Had initiation by Neptune when we crossed Circle in evening. All up to see midnight sun & to celebrate 4th July. Great fun. Bed 1:00am."[10]

The initiation was described in more detail in another expedition member's journal:

> The Captain jumped up excitedly calling, "everyone turn out; there's something the matter"...Captain Johansen then stepped forward and recalled the time of his iniation [sic] and was excused by Neptune. Leffingwell a Chicago University lad was next called for and he presented himself to be iniated [sic]. He was seated on a low chair, after being introduced to Mrs. Neptune, and his face lathered with a white wash brush in a slaty paste. Then shaved with a lathe after which his face was bathed in ice water with a large squirt gun. "Leffingwelly" was game and at the close of the ordeal passed a box of choice Havannas saved especially for the occasion.[11]

All seemed in good spirits. They were excited about the expedition and the possibility of being the first to reach the North Pole.

The next day, Leffingwell noted the scenery was beautiful. He also mentioned that motion pictures of the initiation had been made, but their location is unknown. His reading that evening was *In the Lena Delta,* by George Melville, about the search for Commander G. W. De Long

The *America,* 1901. *Library of Congress, Prints and Photographs Division*

and the *Jeanette* in the Siberian Arctic, and also an account of the 1884 Greely relief expedition.

Leffingwell set high standards for himself in all respects, whether work or athletics. While docked in Tromso, he and one of the expedition doctors, William Verner, also visited the local museum and admired a collection of Arctic birds. The days were spent storing and moving a seemingly endless supply of provisions, taking watches, shoveling coal (referred to as "coaling"), working with the instruments, reading, and taking a few recreational breaks. On July 16, as they were hoisting a whale boat, it fell, crushing the foot of a crew member. Leffingwell assisted with the foot's amputation. The ship left Tromso for Archangel, Russia, that evening.

The *America* reached Archangel, Russia, on July 21, where the expedition obtained 428 dogs (far more than needed and an indication of Baldwin's profligate spending), 15 Siberian ponies and feed of hay and oats, and six Russian men to care for the dogs. The ship sailed on July 25.

Leffingwell shoveled coal to bunkers from the main hatch, and worked with the chronometer on July 28. The next day was even longer, as he and others shifted boxes in the main hold aft so that the bow might rise. They were up until 2:30 a.m. He wrote "securing both anchors on deck and stowing chain. Could have been left till next day."[12] He must have spoken out, for his journal noted that he had been "called down" by Baldwin, who informed him that it was "not mine to say but to do."[13] Leffingwell would continue to express his opinions during the expedition. They departed July 30 for Franz Josef Land.

Leffingwell in Tromso, Norway, 1901. *Deborah Storre Collection*

To Franz Joseph Land

On August 1, the *America* headed northeast into the Barents Sea toward Franz Josef Land. Leffingwell and Mikkelsen began their scientific work, taking weather observations and surface water temperatures every four hours. As the expedition continued northeast, some days were quite foggy, there was heavy snow at times, and the cooler temperatures made handling equipment challenging, especially the sounding apparatus, which measured depth. Leffingwell mentioned the beautiful colors of the ice floes, and in his leisure he read a novel by William Clark Russell, *My Danish Sweetheart*.

The ship entered the pack ice on August 3, after the fog lifted a bit. The temperature was 32°. (Temperatures are given as they appeared in the journals. Leffingwell often failed to indicate F for Fahrenheit or C for Celsius.) They observed two seals on an ice floe. Seas were calm and the entranced Leffingwell felt struck with "Arctic fever." As they continued their journey, they saw bear tracks frequently on the ice floes.

The following day, August 4, the ship's rigging was coated with a half-inch of ice, and snow and sleet fell. It was Sunday, and Leffingwell generally "read service" (perhaps the Bible or the *Book of Common Prayer*) to himself, but on that day also read *Othello*. He made note of their meals: Dinner consisted of macaroni, canned beef, and soup. Supper was dried beef, powdered eggs, cornmeal mush, and applesauce. Expedition food also included canned sweet potatoes, evaporated potatoes, hard tack, rice, beans, pancakes, bacon, and ham. Hunting bear, walrus, and seal supplemented their provisions.

The ship entered the ice pack, then headed southeast, seeking open water or a lead (a narrow opening in the pack ice) closer to Franz Josef Land. Leffingwell was still taking "sights" and by dead reckoning had determined that they had reached latitude 79° 15' north and longitude 42° east and were more than eighty miles from Franz Josef Land.

On August 10, Leffingwell noted that Baldwin went into the "crow's nest" to help navigate the ship through the ice. That evening Leffingwell went for a walk on an ice floe and appreciated being off the ship.

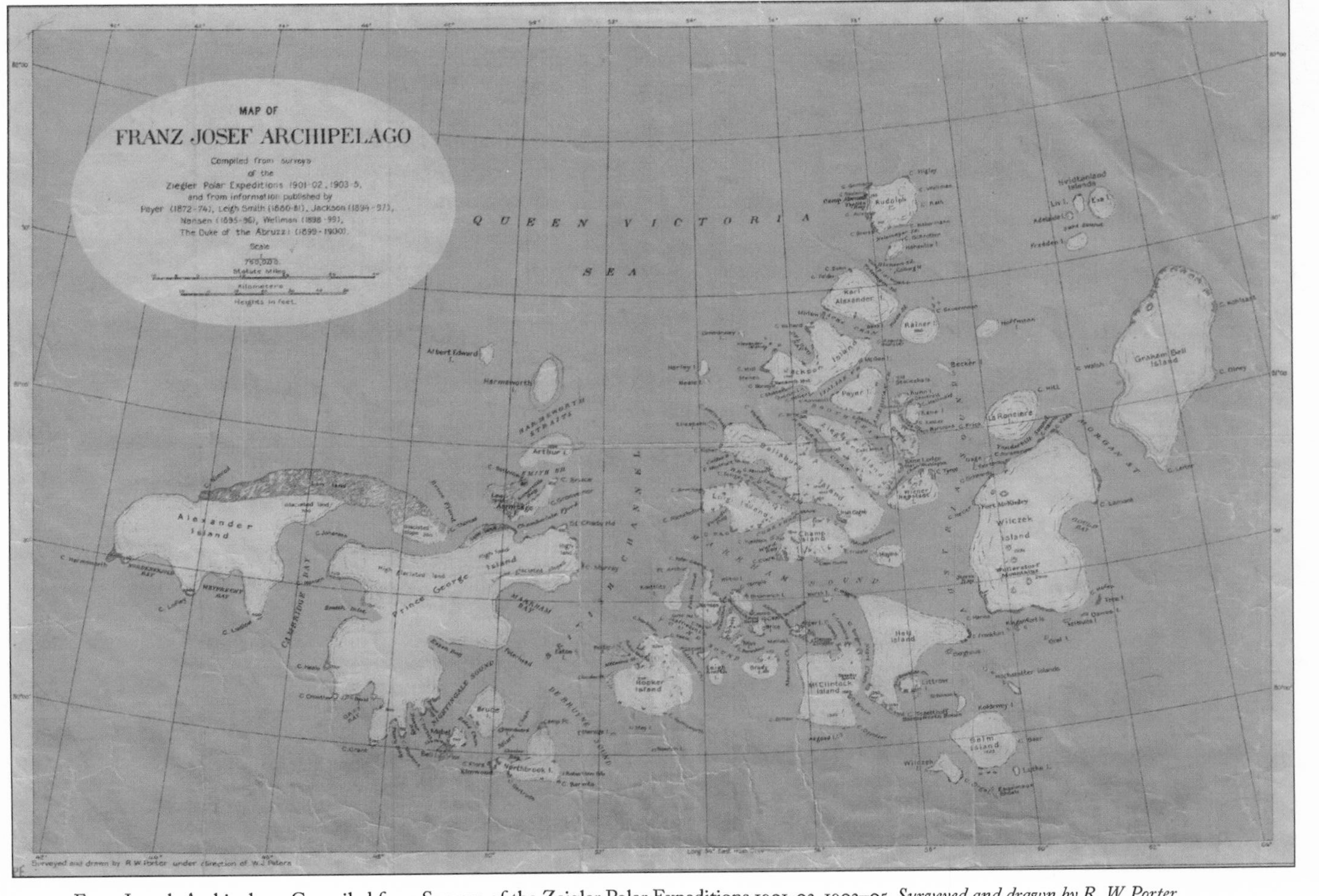

Franz Joseph Archipelago. Compiled from Surveys of the Zeigler Polar Expeditions 1901-02, 1903–05. *Surveyed and drawn by R. W. Porter under direction of W. J. Peters. Andrew B. Graham Co. Lithographers.*

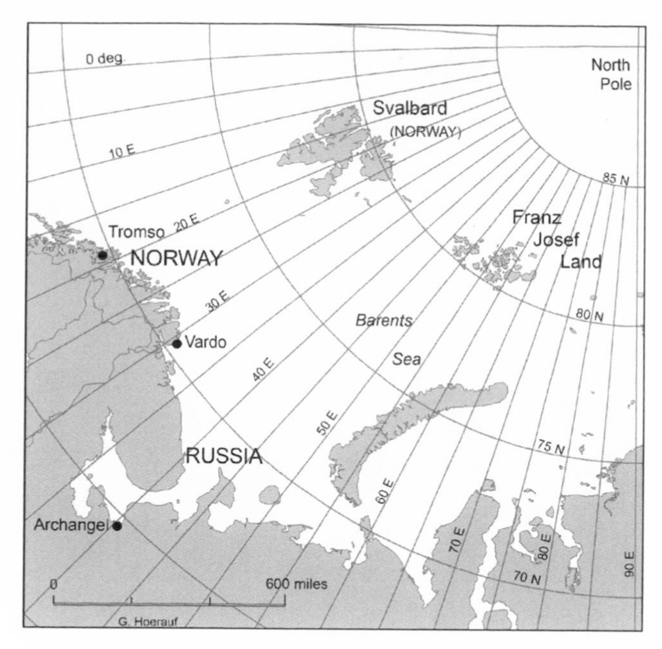

Franz Joseph Land location map. *Eugene A. Hoerauf, Cartographer*

Franz Josef Land

The expedition reached the islands of Franz Josef Land on Friday, August 16, 1901. Leffingwell was on deck as they negotiated the ice floes and made their approach, and immediately recognized Bell Island. The ship anchored against a backdrop of heavy snow and fog at an approximate latitude 79° 54' north and longitude 50° 54' east. As Leffingwell walked on nearby Northbrook Island for the first time, he described the environment: "Poppies out in abundance & saxifrage. Snow buntings. Cries from rookery on cliff mingled with a continuous roar. Beautiful glacier from Mable Isd & N of C. Flora. Low cloud cut off most of view, but enough to make this introduction into our Arctic life a day ever to be remembered."[14]

Two camps were soon established on Alger Island within six miles of each other. West Camp Ziegler, on the southwestern edge, was a depot for provisions. East Camp Ziegler was established on the southeastern edge. The men built several structures to house the dogs, ponies, and animal feed, as well as a weather station. When they were not hauling provisions and the ship was anchored, members slept aboard the *America*. On August 22 Leffingwell, Baldwin, and Porter, the expedition's artist, hiked to the top of a peak about a mile north of the camp. They saw a panorama of open water for many miles, and an opportunity to head north. They hoped to establish a permanent camp on Rudolf Island, located about eighty-five miles to the north, then head for the pole. But the following day, instead of departing, Baldwin retired to his cabin indicating that he was ill. He did not surface again for several days. In the meantime, the accompanying supply ship *Frithjof* departed Camp Ziegler on August 23 for its return to Norway, without encountering any problems getting through the ice. From camp, they observed an increase in open water, but Baldwin remained in his cabin with a sore throat. Over the next few days, Leffingwell's journal noted that Baldwin's sore throat had worsened. Baldwin's strange behavior continued. When Baldwin did surface again on August 28, instead of heading north, the expedition made four trips to the southern islands (over a three week time period), including Hall and Wilczek Islands. Those trips clearly did not contribute to the goal of reaching the North Pole.

The expedition spent September 1 at Cape Dillon on the southwest corner of McClintock Island. Leffingwell's diaries were generally a straightforward recounting of events, but that evening he wrote about the beauty of the Arctic: "Very wild sunset. Colors magnificent," in shades of rose, purple, violet, indigo, cream, and gold. He continued, "All these colors reflected in absolutely motionless & glassy surface of water."[15] Arctic fever again had taken hold.

The ice conditions alternately worsened and improved. On board the *America*, the heavy swell at times required them to keep the engine running because the water was too deep for the ship to anchor. But the same heavy swell broke up the ice which gave them hope to travel north. Throughout the month of September temperatures varied between 30°F. and 15°F, which significantly impacted the ice conditions.

As the days passed, it was not unusual for Leffingwell and the others to spend an entire day coaling, which kept him away from his scientific work. Yet he remained optimistic about attempting the North Pole after talking with Mr. Baldwin, writing "Very fine!! Awake till 3am thinking about it...up 5:30 & shoveled coal all day. Verner, Loth [Mikkelsen] & I get along nicely down there & have fun teaching Loth English college slang. Poker for an hour at night. Wind 36 miles during day."[16] Rather than money, poker was played for sugar rations.

As the expedition ship pushed toward northern latitudes, they reached more ice and tried to ram or "buck" their way through, which burned a tremendous amount of coal and would later impact the length of their stay. They tried to dynamite the ice without much impact. They also tried sawing the ice, which expended the men's energy with little to show for their efforts. A phenomenon known as "water sky" or "dark sky" helped guide them. If the clouds reflected dark, there was water below; if the clouds reflected white, there was ice.

Expedition members tried to understand Baldwin's decisions. In spite of Baldwin's inexplicable emphasis on the unloading and reloading of provisions, they remained optimistic that they would have an opportunity to try for the pole. They were in good humor, as seen in *The Midnight Sun,* an expedition newsletter that appears to have been published only once. One topic was "A Culinary Horror" and included: "Dr. DeBruler, the famous physican [*sic*] and scallion, was very much incensed the other morning on discovering that the chief engineer had used one of his famous flapjacks in repairing a hole in one of his boilers."[17]

After three weeks of trips between Alger Island and the southern Islands, presumably for mapping and observations, they anchored again at West Camp Ziegler.[18] On September 19, Baldwin ordered more provisions to be hauled ashore. His agenda was soon clear.

A Party to Remain Behind

Once Baldwin arrived at Franz Josef Land, he appeared to have difficulty making decisions and was hesitant about proceeding. It is unclear whether he really intended an attempt for the pole once they arrived at Franz Josef Land. He continued to have the men load and unload provisions for no apparent reason. His retreat to his cabin for days when open water was sighted suggested more interest in exploring and mapping Franz Josef Land than traveling north. Perhaps he was simply overly cautious.

Baldwin decided it was time to return to civilization and resupply for another attempt for the North Pole. He announced his decision at tea time on September 21 to leave a party behind for a year of scientific work, indicating that their current latitude was too low to start an attempt north. His rationale was that coal and food reserves would not be adequate to support the attempt. As part of Baldwin's plan, eight men would stay behind to "guard the stores" and continue Leffingwell's scientific work for the next eleven months. That party included Leffingwell, Mikkelsen, Verner, engineer Chas. E. Rilliet, secretary Leon F. Barnard, doctor Chas. L. Seitz, crewman Robert L. Vinyard, and Lucas, who had previously been with the *Frithjof.* Leffingwell and Verner were given leadership roles. According to Barnard's journal, they took about seventy-five dogs to shore and fourteen of the fifteen ponies. Baldwin assigned specific tasks to those remaining behind: additional mapping, photography, construction of two dwellings at the eastern camp, and an "hourly watch for any steamer" between July 20 and September 1.

Provisions were transferred from the ship to shore. "Everything possible" that could be "spared from *America,*" Leffingwell wrote in his journal. On September 27 Leffingwell wrote: "Up 4:30 & landed ponies all am...PM the ice began to come in with S breeze so that east loads had to be abandoned & party to take short leave after a few words of farewell from Mr. Baldwin. *America* left 5:00pm just in time for floe ice came in during night. All happy to have every prospect of passing a pleasant winter here."[19]

The party of men left behind were thrilled to be there. Leffingwell looked forward to finally pursuing his planned scientific work. The party felt comfortable with their provisions, their skills to survive, and were looking forward to their adventure. Baldwin wrote: "Many of us had tears in our eyes when we said goodbye to our comrades, who seemed unusually glad and merry."[20] As the ship hoisted the anchor and prepared to sail away, Baldwin's cousin Barnard wrote, "would be curious to know if any of us ashore, as we watched the well-known process, wished that we were in it. I don't believe any did."[21]

The eight men left on the island spent the next few days improving camp structures, organizing supplies, and dealing with snow and rain. Leffingwell and Dr. Verner helped raise morale. Barnard noted, "I have laughed more today than for some time, notwithstanding the steady, cold drizzle."[22] They called themselves the Aurora Borealis Club. Leffingwell started a new journal, intending to document all of his experiences for future reference, and including his scientific work, his physical condition, his mental state, and his experimentation with clothing.

Their pleasant days were short-lived. On October 2 Barnard wrote, "Day of disappointment and everything else—can hardly jot a word. Loth (Mikkelsen) remarked as he dressed this morning, 'Well, the *America* gets into Tromsoe this morning.'"[23] Shortly after, Barnard reported,

> [The cook, who had stepped outside] stuck his head back in at the door and announced that the 'America' was out there. No one moved—thinking it only a hideous joke in keeping with some of the others that had passed. One by one, however...made his way to the door and looked out; each one returned tho, with such a curious look on his face as to excite and yet mystify the others. I thought all were trying to "rope" in every other one so that the joke would be universal. But I, too, finally went and had my look. There, a few rods out in the bay, and indistinct in the thickly falling snow, lay the ship.[24]

Baldwin's return signaled yet another change of plans: Barnard wrote: "his return here now, therefore, means that we are to accomplish nothing but a more or less complete charting of this archipelago. This day is ten times more dark than the one on which we were turned back before the ice...the spirit is gone out of everybody. On board the air is full of gloom and small talk. I am told that the ship met conditions of ice out in the open which made it seem the better part of valor to return to us."[25]

Baldwin's actual motivation for returning was unclear. Perhaps he questioned his decision to leave a party behind, which would force him to solicit financial backing and return the following year if conditions permitted. If the men did not survive, his reputation would be destroyed. He blamed the return on ice conditions.

Baldwin's plans were equally opaque. Would they attempt to over-winter with limited resources, or would they make the attempt for the pole? Or would they return to Europe?

Leffingwell and Seitz remained at West Camp while the others returned to East Camp and spent the next few days hauling gear there.

Again, Leffingwell found his plans to perform scientific research stymied by other duties. Baldwin ordered him to come to East Camp Ziegler on October 5, bringing with him the ponies, dogs, remaining walrus meat, and some pipe material. The next day Baldwin changed his mind once again and decided to leave a party behind composed of the same group of men. Barnard wrote that they were pleased with the new orders. It was not to be.

Ten days later Baldwin wrote with further instructions, none of them concerning the scientific work. Baldwin had the expedition members moving gear from location to location, and resentment grew. Barnard wrote, "together with the re-handling, for the 127th time, of a liberal part of it only last Sunday. And yet there is no voice raised in serious complaint against this thing that has continued so long; every American up here is here in the supreme hope of, and prayer for, every success to the Expedition. Everyone feels that if we cannot reach the pole, then would we see this archipelago successfully and accurately charted."[26]

In additional to the planned ambitious scientific work, Leffingwell had his sights set on the North Pole. He did not know when the "final dash" would happen, but he wanted to be prepared. Later the same day, however, he learned that the ship had gone north again, so the "final dash" was not imminent. The *America* returned to West Camp on October 12.

Leffingwell remained at East Camp along with Seitz. He spent a significant amount of time caring for the dogs and preparing loads for transport to West Camp. He made observations as the weather allowed. October 18 came and went with the last sighting of the sun until late February. Temperatures ranged from 30°F to -20°F, and Leffingwell had his first experience with frostbite.

Winter in Camp

Between their other chores, Leffingwell and Mikkelsen continued work on the astronomical observatory. The observatory was an important part of the scientific work, allowing them to determine the time and their location, which in turn would allow Leffingwell to map the region accurately. They added a canvas roof to the structure. During early November temperatures dropped to -20°F and -30°F during the day. By mid-November temperatures were consistently below zero. The men ran telephone lines between the America and West Camp, including Leffingwell's observatory, and electric lines and lamps on board ship and at West Camp.[27]

By mid-November the only natural light occurred between 10:00 a.m. and 2:00 p.m. Lack of visibility created angst when several expedition members working outdoors discovered a bear only a few yards distant. They ran, chased by the bear which finally turned away. From then on, everyone carried a gun.

Leffingwell's scientific work was again secondary to other chores assigned by Baldwin. For two days after Thanksgiving, Leffingwell helped haul ice to the ship from a nearby iceberg and then spent four days cooking mush for the dogs.

On December 6, 1901, a major storm hit with gale force winds and snow. It relented about noon on December 8. Finally able to feed the dogs, Leffingwell spent eleven hours cooking mush for them.

Not surprisingly, the astronomical observatory was half-full of snow with one wall gone. They dug it out, rebuilt, and resumed their work. Leffingwell spent evenings working on sketches of Franz Josef Land and mentioned the deficiencies of using smoky whale oil lamps for light. His eyes were suffering from the smoke, and he had lost his glasses. He also worked on overhauling instruments, including the magnetometer.

Christmas was low key. Most dressed up for the occasion and a dining table was decorated. The evening of December 31, the men celebrated the New Year with cigars and a variety of musical instruments for entertainment.

On January 1, 1902, Leffingwell wrote "Up 9:00 & started to celebrate with a new suit of underwear but had to hang them over stove to get ice off."[28] Baldwin spoke that evening about their accomplishments over the previous year, but provided no details or plan for reaching the North Pole.

Barnard wrote about Mikkelsen again on January 4:

> Loth seems to have gone daft on bets. The fellows dare him to the most ridiculous things, and he takes them up just to show that he can. Today he capped the climax...The bet was that he could not go barefoot over to the camp from the ship and back again. He hardly hesitated a second and astonished all by taking the bet and meeting the ration of sugar and butter. We all tried to stop him but he was set upon doing it, and so after coffee he and a half dozen others went out on the driveway approach and he took off his stockings and was off. Two of the fellows went with him with his boots etc., so as to be on hand in case of an accident, but he returned all right and we too [sic] him into the galley and chaffed his white feet until they glowed. He was triumphant and immediately called for his sugar, etc.[29]

Leffingwell continued with his observations, improvising and problem solving when equipment broke or malfunctioned due to the harsh conditions. His eyes were again sore from smoke. With Mikkelsen he remained optimistic about a final dash for the pole.

The men were busy aboard ship constructing sledges, sleeping bags, tents, sails, and clothing, yet another example of an ill-prepared and disorganized expedition that jeopardized the safety of its members. Clothing considerations had always been of concern. In preparation for the expedition, Baldwin had written to its members, "You will be expected to provide yourself with necessary ordinary underwear and plain clothing, all special Arctic gear such as fur suits, boots, etc., will be furnished free."[30] The reality proved to be quite different. Their clothing included wool clothes, wool hats, wool mittens, and canvas for warmth. Canvas was used for wind coats and to cover woolen mittens. The wind coats kept them warm to -30°F.

There were frequent trips, often quite dangerous, between East and West Camps hauling loads and supplies. They were not always successful. Weather conditions were brutal. Cold temperatures affected everything they attempted to accomplish. During the month of January Leffingwell recorded temperatures to -42°F several times. The wind chill made conditions far colder. Based on conversations with Baldwin, the men still anticipated an attempt for the North Pole.

On January 27 Leffingwell and three others took several sledges, numerous dogs, and supplies from the ship to West Camp. They reached it in about two hours, but during their return, and in darkness, they were caught in a nasty storm. All were struggling with the dogs and having problems with the harnesses.

Leffingwell decided to abandon the sledges and travel by foot to the ship. They sighted the light of the ship, but traveling the remaining two miles took close to three hours. Leffingwell described this dangerous trip and the effects of hypothermia, which he attributed to "lack of exercise." He later wrote: "In my own case, I was in a physical state absolutely novel to me while not feeling tired in any muscle I found myself going to sleep on the sledge when during stops. My balance was easily lost and I kept falling down at every irregularity. I feel much chagrined at this but knew before starting that I had never in my life before been in such a poor physical condition from lack of exercise. Andree says the same of himself."[31]

On February 7, not six months after arriving in Franz Josef Land, Baldwin again wrote to Leffingwell and ordered him to pack up the scientific equipment for eventual shipment. There would be no further scientific work accomplished. Baldwin explained that the upcoming sledge travel north would not allow time for the scientific work. Leffingwell's journal makes no reference to these decisions or instructions, but he did not immediately comply.

While in preparation for the trip north, Leffingwell read. His latest book was Sir George Back's narrative about an early 1830s expedition down the Firth River and along the coast of the Arctic Ocean in today's Yukon Territory.

The next week Baldwin gave a brief outline of plans to head north. He had private conversations with Leffingwell and Porter, on which Leffingwell did not elaborate in his journal. On February 17 Baldwin announced that their destination would be Rudolf Land [Island]. There would be no attempt to reach the North Pole. Weather delayed their departure. A week passed.

By February 25 it was clear that Baldwin and Leffingwell were at great odds. There was a confrontational exchange between the two over Leffingwell's work plan for the day. Baldwin ordered Leffingwell to stop his scientific research and write a report summarizing his accomplishments. Circumstances would determine whether he continued any research.

On March 3, 1902, Leffingwell delivered a nine-page report that detailed hydrographical, meteorological, astronomical, ornithological, magnetic, and topographical surveys, and included a list of equipment. The hydrographic work consisted of ninety soundings within the archipelago. Surface water temperatures and density had been recorded during a two-week period in August 1901. Also recorded were thirty soundings, with temperature and density, and bottom specimens. Meteorological work included use of a bulb thermometer, barometer, and recording of clouds and wind. Astronomical observations determined latitude and longitude at both West Camp and East Camp, while magnetic work measured the amount of variation from the zero position of an instrument. Topographical work consisted of approximately twenty each of sketch maps, profiles, and views. Leffingwell took angles using an altazimuth from nine stations and used a sextant from six stations.

Leffingwell's report included the impact of the cold on the various instruments, and the challenges of adjusting them to provide accurate readings. Given his ambitious plan for scientific work generated before the expedition sailed, Leffingwell later expressed disappointment in the work accomplished.

The Sledge Trip North

Two weeks had passed since Baldwin's announcement, but the expedition had not yet started their trip north. Leffingwell's last journal entry is dated March 2. More than six months after arriving in Franz Josef Land, Baldwin announced on March 6 that they would head north towards Rudolf Land [Island].

As they prepared for the sledge trip north, Barnard wrote on March 7 that everyone was in high spirits but, "We are disappointed. We did not come up here to explore and map but simply and only to discover the North Pole, nor did we come up so as to make up a story, as some put it...And, from another point of view, it is necessary that we do not stop at charting. Baldwin and Mr. Ziegler have called the attention of the world upon the expedition and we must not be found wanting."[32] Their reputations were on the line and all of them recognized it.

March 15 found the men moving loads of goods from one cache to another further north. They were scattered about and from time to time would regroup. The weather was not cooperating and Barnard wrote:

> They lost the trail and we were in for it. The wind was blowing about 20 to 25 miles but to our backs. My cap was frozen stiff and my chin like a cake of ice...We were tired. Baldwin was excited and surprised me very much. Leffingwell was right in his suggestions as to where we were and had Baldwin not followed them we would not be here tonight. We wandered until ten o'clock and found Alger Island which Baldwin was not sure of for a long time. Baldwin finally disappeared entirely and the party had no leader and no one any authority. Leffingwell picked out the direction but no one would go that way for some time, i.e. after we reached Alger Island, and when we did do so the light of the ship soon appeared right ahead.[33]

Things were deteriorating and, as Barnard recorded, in the evening of March 16 Baldwin called a meeting. "Leffingwell promptly asked if there would be sleeping bags taken along on every trip after we get out on the trail. Baldwin said this would not be granted at all. Leffingwell asked to be relieved from the sledging trips."[34] According to Barnard, Baldwin's plan was to move supplies in a series of stations separated by five to ten miles, which Baldwin felt would not require sleeping bags. Though he later relented, Baldwin continued to lose the respect of his men.

Barnard's journal ends on March 17, 1902. There must have been a reason that both Leffingwell and Barnard stopped writing. Leffingwell may have been concerned that his journals would become public. Barnard had openly expressed his thoughts about the expedition, but switched his later journal entries to shorthand to maintain his privacy. Leffingwell's journal, clinical and innocuous, seldom reflected emotion. Barnard's journals provide the best insight into daily life and personal struggles during the expedition.

Baldwin dispatched Leffingwell and others with sledges to a separate depot to transport and then rejoin the main party. Baldwin traveled by pony and when he reached them he instructed them to add more to their loads. Wind and snow increased, causing the sledges to struggle

Sled trip north from Camp Ziegler, 1902. *Library of Congress, Prints and Photographs Division*

through often waist-high drifts. Baldwin was over one eighth of a mile away and appeared oblivious to their dire situation. The procession came to a stop in whiteout conditions and the expedition members assumed Baldwin would have them settle into tents for the night. Instead he ordered some loads shifted to other sledges and returned to his pony to resume travel. Barnard responded to Baldwin in an uncharacteristic way—direct and assertive about the conditions and the men and sledges. Baldwin then ordered the men to pitch tents. They had been inside their tents about a half hour when Baldwin reappeared. He announced that the winds were dying down and they should prepare for travel. The men were incredulous at Baldwin's behavior.

By April 17 all supplies, 160 dogs (out of the original 428), 15 ponies, and 27 men arrived at Kane Lodge on Greely Island, a direct line distance of approximately 15 miles from their previous camp. The ponies hauled sledge loads estimated by Baldwin to be between 800 and 900 pounds. On April 22 Leffingwell and Baldwin were taking bearings on islands to the north, including Wiener Neustadt Island and Whitney Island (known today as La Ronciere Island). Baldwin planned to continue travel north on Greely Island.

They traveled north between Stelicza and Becker Islands to the west point of Rainier Island late on April 24. Leffingwell and Baldwin set out for the summit of Rainier Island on April 26, arriving at 9:30 p.m. When the expedition reached Coburg Island, Baldwin and Leffingwell made an ascent to obtain more angles to be used in mapping the area.

On May 1 Leffingwell determined that they had reached a latitude of 81°33' North. They traveled north toward Cape Brorok and Cape Auk on the west coast of Rudolf Island, arriving on May 3. From Cape Auk, Baldwin looked to the north and saw open water in Teplitz Bay. Baldwin had determined (without sharing it with the rest of the expedition) that the following year he would return and travel along the east coast of Rudolf Island to catch the northward drift of the pack ice and attempt the pole.

They decided to establish their northern depot at latitude 81°45' north, intending to utilize it the following year. By then Baldwin and Leffingwell were working and traveling together. In spite of their differences, it is clear that Baldwin needed and respected Leffingwell's abilities. If Baldwin did not intend to attempt the pole, then it was important to have something show for their efforts. Along with depositing caches for a future expedition, completion of the mapping of Franz Josef Land would have to suffice.

On their return south Baldwin and Leffingwell took a side trip, which included a stop at Hohenlohe Island where Leffingwell took more angles for mapping. Baldwin then announced plans on May 10 for Leffingwell and a party to travel to Wilczek Land, approximately twenty miles east, to further survey and complement Baldwin's work from 1899. Baldwin later learned that the survey was not completed; Leffingwell had taken a bad fall and was recovering back in camp.

On May 11, Baldwin described a channel: "noting the appearance of Rainier Island to the northeast, through the northeastern exit of the channel, and gave to it the name of Leffingwell channel, in part recognition of Mr. Leffingwell's willing cooperation in the geodetic work of the expedition and of which the many observations made by his bear testimony."[35]

The expedition continued their journey; Baldwin named Ziegler Island in honor of the financial backer of the expedition. On May 14, the small party located the cabin that Nansen and Johansen had built and occupied between August 1895 and May 1896. They left Kane Lodge on Greely Island for the last time five days later, and the expedition headed south again to Alger Island and the *America*. When they saw the masts of the ship on May 20, there was much cheering.

The *America* left Camp Ziegler on Alger Island on June 30 for the return to Tromso. On July 1, 1902, the *Frithjof* departed Norway for Franz Josef Land as a relief expedition to the *America*. The *America* arrived in Norway in early August, but it was early September before the *Frithjof* returned. Baldwin blamed the expedition failure on the ice conditions, coal expended, and lack of reserves. But his troubles did not end with the expedition's return to Norway.

Expedition Controversy

Controversy over the expedition's organization, preparation, and intent had begun before the *America* set sail. Well-funded, the expedition nonetheless appeared ill-prepared for traveling in the Arctic. Lack of clothing and shelter were two of the best examples. Baldwin was circumspect about his plans, but his behavior suggested that the intent was not to make an attempt for the pole, but to map Franz Josef Land and establish caches for a future expedition. Baldwin's decision to leave a party behind and have them sign new contracts under his control was the decision that would spark the greatest controversy. Baldwin appeared to be very concerned about his professional image and that of the expedition. He had confrontations with expedition members and was unsure of how their expedition would be regarded.

As a result, Baldwin issued another order to expedition members which required their acknowledgment, compliance, and signature while at sea on July 25, 1902. It required accounting and relinquishing possession of all items to Baldwin before being discharged from the expedition. Following the return of the expedition, the relationship between Baldwin and Ziegler quickly dissolved into a battle played out in the press. It was a long, bitter, and contentious dispute over Baldwin's questionable decisions and actions during the expedition, and the new contracts he had required the men to sign who were to be left behind for a year. Baldwin cited the lack of coal reserves as a significant reason for not attaining the pole. But the shortage of coal had also placed expedition safety in jeopardy.

Ziegler was part owner of the *Brooklyn Daily Eagle*, one of the newspapers that published the statements of many expedition members. At Zeigler's request, expedition members had written to his secretary, Mr. Champ, and it is likely that their statements were published with their knowledge or permission. Almost all of the statements were scathing, including the one written by Leffingwell.

Leffingwell's letter to Champ was written November 22, 1902. Statements by other expedition members help explain Leffingwell's wrath. His were not isolated sentiments. Like other expedition members, he was an idealist, frustrated and disappointed with the expedition from beginning to end. Leffingwell was also a confident young man who had left his graduate studies to perform scientific studies and take part in an attempt to reach the North Pole. Leffingwell's letter was eventually published in the *Daily Eagle* on December 27, 1903, along with statements from other expedition members. One of the most legitimate points that Leffingwell made concerned the lack of opportunity to conduct the scientific studies he had been hired to perform. His statement detailed frustration and disappointment in all aspects: the dearth of clothing and provisions, lack of cooking stoves on trips, controversy over allowing sleeping bags to be carried on sledge trips, and Baldwin's lack of leadership skills.

In their responses, the other expedition members questioned Baldwin's judgment, his integrity, and his apparent stronger interest in mapping Franz

Josef Land than making the pole. They also cited the lack of preparation, the use of tin cans for cooking utensils, and the incessant loading and reloading of cargo between ship and shore.

Carl Sandin, a Norwegian expedition member, wrote a lengthy letter to Baldwin, counter to the other critical letters, to express his support and praise for Baldwin's actions and behavior.

In spite of the negative media attention, Baldwin attempted to garner support for another polar expedition. On October 15, 1902, Leffingwell received a telegram from Baldwin indicating that Mikkelsen wanted to go the following year if Leffingwell did. Leffingwell knew that was not true. Baldwin had hoped that Ziegler would sponsor another attempt to reach the pole, but the relationship deteriorated further and Baldwin never returned to the Arctic. He eventually worked as the historian for the Office of Naval Records and Library in Washington, DC, and died in an automobile accident in 1933.

A year after the Baldwin-Ziegler Expedition, Ziegler financed another expedition to include an attempt for the North Pole. Known as the Fiala-Ziegler Expedition, it sailed under the leadership of Anthony Fiala, photographer on the first expedition. Eventually Leffingwell made application and was once again offered the chief scientific officer position. Without assurances that he would be allowed to do the scientific work that had been planned for the previous expedition, he declined the offer. The expedition also failed in its attempt to reach the North Pole.

Though disappointed that the expedition had failed in many ways, Leffingwell's accomplishments included data from scientific measurements and his map of Franz Josef Land. There were many other positive outcomes that influenced the rest of his life. The expedition provided Leffingwell with invaluable training and Arctic experience. As chief scientific officer, his earlier training in instruments would serve him well in his future endeavors. He learned about expedition leadership, organization, and interactions with others. And he learned about survival in the harsh Arctic climate: about food, clothing, sledges, and travel. Most of all, the expedition also provided him with confidence in his abilities and a strong desire to return to the Arctic. Leffingwell along with his friend, Ejnar Mikkelsen of Denmark, were determined to organize their own expedition and travel to one of the few remaining unexplored areas, north of Alaska, in search of land. Determination and resiliency would be critical to their success.

CHAPTER 2

Interlude: 1902–1906

Leffingwell led a busy and productive life following his return from the Baldwin-Ziegler Expedition. He continued his studies at the University of Chicago earning consistently good grades in geology. His course work was enhanced by summer field seasons examining glacial processes and surficial geology under Dr. Rollin D. Salisbury.

He spent the summer of 1903 in the Big Horn Mountains of Wyoming. That fall, he returned to Knoxville, Illinois, to serve as superintendent of his father's St. Alban's School. In addition to overseeing the operations and curriculum of the school, Leffingwell coordinated weekly camping trips.

Other exploration invitations arrived. Earlier in 1903 he had received an invitation from Frederick Cook to explore and climb Mount McKinley (now named Denali) in Alaska. He declined the invitation, but indicated that he might be able to go the following year. Leffingwell had other goals in mind.

His desire to return to the Arctic was profound. In December of 1903 Leffingwell wrote to Baldwin in an attempt to reconcile some of their differences, and because he had another agenda. Leffingwell had turned over his journal and map to Baldwin as ordered and indicated that it was currently in the hands of Ziegler. Leffingwell had requested Ziegler return the items several times, but had not received a response. He considered suing Ziegler for their return. Leffingwell's hopes of getting Baldwin's endorsement were dashed on December 27, when Ziegler published Leffingwell's condemning statements about Baldwin.

"I am very anxious to go north again," Leffingwell wrote to Baldwin. "You understand the feeling. I did not like to go under a man like Fiala, who had no more experience than I (not so much) and who did not show up especially well in the field; but I saw that if I did not go on their conditions I would loose [*sic*] everything of the first year and be ignored when the Ziegler Expdn's report was publish[ed]. Ejnar Loth [Mikkelsen] and I are trying to get up a trip to explore to cost [*sic*] North of Shannon Island.[1]

Shannon Island in eastern Greenland was the location of the storage depot Baldwin had established for the Baldwin-Ziegler Expedition. Leffingwell and Mikkelsen were hoping to use the provisions, but were unaware Ziegler had already promised them to Commander Peary in his bid for the North Pole.

Leffingwell then had another expedition in mind, and asked Baldwin for a letter of recommendation:

> I am trying to go with Peary next summer. If you will write him a recommendation of me, telling of the various positions of trust I held under you and of the satisfactory account I gave of myself: Of my eagerness for such scientific work as there was time for and of my ability along different lines; it will go a long way toward settling the various misunderstandings and differences we may have had. If you will do this and send me a copy, I will consider it the best evidence of your good will that you could afford me. If you seriously contemplate publishing a narrative, perhaps I could furnish data from memory on points you [may] have forgotten about.[2]

Leffingwell remained hopeful that Baldwin would support him in his efforts to become a member of Peary's expedition to the pole. Leffingwell sought recommendations from several other members of the Baldwin-Ziegler Expedition as well. Archibald Dickson and J. Knowles Hare were two of them; Leon Barnard was another.

Peary responded to Leffingwell's inquiry in December 1903, saying he would consider Leffingwell's request to become a member of his polar expedition but could not provide an opportunity for Leffingwell to conduct his scientific work. Peary's only goal was to be first to the North Pole. In brief correspondence dated January 29, 1905, Peary dismissed Leffingwell, indicating that he was too busy getting the ship ready to deal with personnel matters and that he hoped to sail in March. Peary may well have read about the Baldwin-Ziegler expedition controversies in the press.

Leffingwell wrote of the Baldwin-Ziegler expedition in the St. Alban's School publication *File Closer*, noting that the "reason for our return as being foolish to start on a 500 mile race with a 100 mile handicap; and also as lack of coal & provisions etc."[3]

Leffingwell indicated to Baldwin that he was "anxious to get some credit for the work I personally did, especially that the map I made be published before Fiala comes back and gives it out as his."[4] Leffingwell had mapped the area as best he could from the ship and excursions on land, and eventually turned the map over to Ziegler (see page 24). Leffingwell's map of Franz Josef Land was apparently published after the return of the second Ziegler Expedition without credit. In a surprising gesture, he offered to give Baldwin a map drawn from memory to publish. "I could patch up a map which would be good enough to print. If you care to publish this, it is at your disposal. I think that if you had trusted me a little and had allowed me a few photos, a copy of the map, my diary, etc. you now would be better off yourself."[5]

Leffingwell's desire to return to the Arctic was temporarily unfulfilled, so in summer 1904, he traveled to the Leadville, Colorado, area for more geology field work under the guidance of Dr. Salisbury.

But his experience in the Arctic led to other opportunities. In 1904 he was a founding member of The Explorers Club of New York City. In 1905 Leffingwell's summer field work focused on the Lake Chelan area in Washington State (where his sister, Hortense, was then living with her family). His plan was to write his dissertation on "The Glaciation of Lake Chelan, Washington." Following his field work, Leffingwell returned to the University of Chicago and worked as a research assistant in the geology department.

Among the requirements for a PhD were language exams. Leffingwell had always had an affinity for languages, so it was no surprise that in January 1906 he passed his examinations in French and German. A month later, he passed his physics final examination. Leffingwell was approved for candidacy for a PhD in geology and physics on January 27, 1906, following completion of coursework. His dissertation was never completed.

But he was still determined to return to the Arctic. His experiences under Baldwin had strongly motivated him and his good friend, Ejnar Mikkelsen, to organize their own expedition. Leffingwell and Mikkelsen decided to jointly command an expedition in 1906. They first needed to secure financial resources.

CHAPTER 3

The Anglo-American Polar Expedition

At the conclusion of the Baldwin-Ziegler Expedition, Leffingwell and Mikkelsen agreed they would each contribute $5,000 toward their own expedition ($125,000 in today's dollars) before they could move forward. The goal of their Anglo-American Polar Expedition (AAPE) was not the North Pole, it was science. And the question of the day was whether there was land north of Alaska. It was an ongoing debate among scientists, geographers, and explorers at conferences and in newspapers. Heeding Fridtjof Nansen's 1897 call for contributions to knowledge through science, Leffingwell and Mikkelsen would seek to resolve that debate.

The movement of tides and currents in the Arctic Ocean, as documented by Rollin A. Harris of the U.S. Coast and Geodetic Survey in 1904, suggested a land mass north of Alaska.[1] Others, including the president of the Royal Geographical Society, Sir Clements Markham, also believed in the presence of land. If land was found, Leffingwell hoped that the discoverer would call it "Harris Land" after Rollin Harris.

The Anglo-American Polar Expedition would travel to the west coast of Banks Island in the Canadian Arctic, northeast of the Alaska-Yukon boundary, and follow the pack ice in search of the continental shelf by collecting soundings, or depth measurements. If the continental shelf was close to Banks Island or the Alaskan coastline, land likely did not exist. If the shelf lay hundreds of miles to the north or west, an undiscovered land mass might be found. The isolated area was virtually unexplored. Mikkelsen felt that any pole attempt from northern Alaska was unlikely. But aside from attempts for the North Pole, the area was wide open for scientific study.

The Beginnings

Leffingwell's initial financial contribution was provided by his wealthy parents. His father was unsurpassed in championing the younger

Leffingwell's interests. For Mikkelsen it was the opposite. Mikkelsen personally contributed $1,500 and with enormous determination and effort, came up with his $5,000 contribution. Most of the expedition details fell to Mikkelsen as Leffingwell was involved in geological fieldwork and university course work during 1905 and 1906.

From the beginning, Leffingwell and Mikkelsen had agreed to be joint commanders. Ejnar Ditlevsen, who had been with Mikkelsen on the 1900 Georg Amdrup expedition to eastern Greenland, was asked to join as artist and zoologist.

Leffingwell invited Harvard University anthropologist Vilhjalmur Stefansson to Chicago to discuss the expedition and then asked him to join, fulfilling the requirement of one contributor who donated funds contingent upon having an anthropologist included on the team. Stefansson, age 28, would serve as ethnologist, and it was agreed that he would travel via the Mackenzie River and meet the expedition at Herschel Island near the river's mouth. It would be his first trip to the Arctic. The expedition accounts show that Stefansson was provided with at least $275 from the AAPE for his expenses. Additional funds were provided to him by Harvard.

Dr. George Howe was the fifth and final member to join the scientific staff as expedition's surgeon. He had graduated from the Harvard University Medical School in 1904. Stefansson later wrote that Howe was also an anthropologist.

The Plan

After purchasing a ship, the team planned first to sail to Kodiak Island where they would try to procure a large grizzly bear for the Tring Zoological Museum in England, at the request of a major expedition donor, Baron Walter Rothschild. The expedition would then continue on to Siberia to purchase sixty sledge dogs and a pony or two for use on the ice. Ponies had been used on the Baldwin-Ziegler expedition and by other earlier expeditions.

From Siberia, they would travel along the north coast of Alaska to the mouth of the Mackenzie River and rendezvous with Vilhjalmur Stefansson about August 20. En route, they planned to conduct tidal observations along the Alaskan coastline between Harrison Bay and Herschel Island. Once united with Stefansson, they would travel to

Prince Albert Land east of the Mackenzie River and establish a cache for provisions. From there, the plan was to overwinter on the northwest side of Banks Island at Minto Inlet.

Leffingwell and Mikkelsen intended to travel northward from Banks Island over the pack ice accompanied by one other man, while the ship returned to Victoria, British Columbia. If they located the edge of the continental shelf, they planned to follow its edge heading westward. They established a latitude of 76° 30' north and a longitude of 150° west as their goal for making further decisions. If the edge of the continental shelf was not found, they would decide whether to continue on west to Wrangel Island north of Siberia or turn toward the Alaskan coastline. Their contingency plan depended upon being picked up by a whaling ship, or living with the Inupiat until they could make their way out in 1908.

A copy of the plan was forwarded from Ottawa to the North-West Mounted Police (later the RCMP) dated March 30, 1906. Mikkelsen wrote:

> The expected results of the Anglo-American Polar Expedition are in short as follows:
>
> Tidal observations along Alaska and Banks Island.
>
> Geological, Ethnographical, Zoological collections on western Parry Islands
>
> Meteorological observations during a two years' stay.
>
> Partly exploring the interior of the western Parry Islands.
>
> Line of soundings out from Pr. Patrick Island.[2]

As news spread about the expedition in the press and journals, so too did speculation about their motives. Some articles suggested that gold deposits might be found in the newly discovered land. Or perhaps a route to the North Pole might open. Sovereignty was a huge issue. If they did find land, which country's flag would be hoisted? A governor general of East Siberia declared that they would ensure that no American flag would be raised if the expedition found land near Wrangel Island, north of Siberia. It was also suggested that the AAPE hoped to learn more about Robert McClure's ship, the HMS *Investigator,* which had been abandoned in 1853 during the search for the Franklin expedition.[3] As members of the AAPE later learned, press reports and rumors would plague their families with doubt and worry.

Funding the Expedition

The role and significance of the American Geographical Society (AGS) and the Royal Geographical Society in London (RGS) in their ongoing support and encouragement to Mikkelsen and the goals of the expedition cannot be emphasized enough. These organizations and their members offered financial support time and time again to the AAPE. The organizations themselves were not wealthy but their individual members were very supportive of exploration efforts.

Mikkelsen's reputation and credentials were solid as a result of his experience on the Amdrup Expedition. In February 1903 Mikkelsen approached Sir Clement Markham, the president of the Royal Geographical Society, about possible funding if Mikkelsen could not obtain it from Denmark. Markham was very interested, having previously written on the topic of land north of Alaska. Mikkelsen again approached the RGS in November 1905, after one of his Danish financial supporters dropped out. Mikkelsen also initiated contact with the publisher Heinemann of London for a book contract and advance. When Mikkelsen promoted their expedition to interested parties, he not only addressed the tidal studies that suggested land north of Alaska, but also focused on scientific opportunities to learn more about currents, winds, and the flight paths of migratory birds.

J. Scott Keltie, RGS secretary, was tireless in his support of Mikkelsen and the expedition. He raised funds through the "friends" of the RGS and met with the publisher, Heinemann, in December 1905. Keltie continued to offer support, guidance, and advice to Mikkelsen throughout the following two years, even suggesting that their vessel be named after the Duchess of Bedford, a major RGS contributor. By the end of the month, Mikkelsen had signed an agreement with Heinemann. However, one of the requirements of the agreement was that Mikkelsen be identified as the "sole" commander of the expedition. This understandably did not set well with Leffingwell but the two resolved the issue during their journey north, and Mikkelsen consistently referred to Leffingwell and himself as joint commanders.

At an RGS meeting in February 1906, Mikkelsen presented the plans he and Leffingwell had formulated for the expedition. One of the attendees was Fridtjof Nansen, who had inspired Leffingwell's original interest in Arctic exploration. Nansen offered his "approval and support" for the expedition.[4] Through the RGS, the Duchess of Bedford responded with the first of several donations. Mikkelsen also contacted and met with the AGS in New York in early 1906, referring to the expedition as a joint venture with Leffingwell. In his proposal to AGS, plans had changed. Leffingwell and Ditlevsen would travel down the Mackenzie while Mikkelsen would travel north on a small ship. They would meet at the mouth of the river and travel to Banks Island to overwinter, then conduct the planned field work. Mikkelsen provided the AGS with a detailed estimate of food consumption for men, dogs, and a horse, as well as sledge weights.

Mikkelsen worked diligently and tirelessly at promoting and justifying the expedition. He was twenty-six years old, five feet nine inches tall, and about 160 pounds, with seemingly boundless energy. Contributions ranged from large sums from a Who's Who list of wealthy international donors, to small sums from the interested public. An outpouring of financial support came in from Canadians, Danes, and Americans.

Initially the Royal Geographical Society and American Geographical Society and their "friends" groups supported the expedition with $1,000 and $3,250 respectively and later provided further support. The Duchess of Bedford repeatedly provided funds totaling over $1,250. Lord Rothschild donated over $1,500 in several installments. Heinemann, the publisher, advanced $3,000. According to Mikkelsen, John D. Rockefeller provided $5,000, and other financial supporters included Alexander Graham Bell and the Carnegie Foundation.[5]

Mikkelsen sought funds in Victoria, BC, in early May 1906, then traveled to Seattle to speak before Danish citizens there, raising $500 in three hours. As a result of public events and newspaper articles, the Canadian government and Canadian dignitaries donated significant funds. Many individuals contributed funds ranging from $5 to $25.

The AGS committee approved $1,000 funding, and four AGS "friends" offered an additional $2,000 that would provide the funds to purchase a vessel. In exchange, the AGS and RGS would receive all scientific results of the expedition "for publication simultaneously in their respective magazines," and the book to be published with Heinemann would "state that the expedition was made under the auspices of the Royal Geographical Society and the American Geographical Society."[6]

The AGS also required progress reports from the Arctic and speaking engagements upon their return.

By the time they sailed on May 20, 1906, they had raised (and spent) between $22,000 and $26,400 (around $600,000 today).[7] Leffingwell's father contributed about $8,000 of the total. Non-monetary contributions included scientific equipment, food, and dog harnesses.

A Ship and a Flag

Mikkelsen still needed help overcoming one crucial problem before sailing—his nationality. Maritime law specified that a ship could not sail from U.S. or Canadian waters under that nation's flag unless the owner was a citizen. The persistent Mikkelsen, a Danish citizen, went straight to the president of the United States. With the assistance of the Danish Consul, he met with President Theodore Roosevelt on February 10, 1906, to enlist his support for the expedition and to ask if Congress might assist by passing legislation providing an exception to the law.[8] Roosevelt turned the matter over to a Senate Committee. According to Mikkelsen, Roosevelt offered the assistance of the U.S. Revenue Cutter Service, predecessor of the Coast Guard and responsible for upholding the law on U.S. waterways. Roosevelt also indicated that he would have U.S. Customs process their provisions without opening and without delay. The matter of the ship and its flag was unresolved.

After meeting with Mikkelsen on February 23, Chandler Robbins, AGS Domestic Corresponding Secretary, wrote Senator Henry Cabot Lodge, a Fellow of the AGS, requesting assistance. Mikkelsen wanted to purchase an American vessel and take it north from San Francisco. "He is told that a special act of Congress is required to permit him to do so."[9] Robbins wrote President Roosevelt on March 3. "Drafts of a bill to accomplish the purpose were forwarded to the Chairman" of several committees and the House of Representatives.[10] However, in a letter to Mikkelsen on March 10, Robbins expressed his doubt that such an act would be passed. And it was not. Regardless, Mikkelsen did not find a suitable vessel in San Francisco; they were all too expensive.

The option to travel north by whaling ship disappeared because eleven ships were caught in the pack ice off northern Alaska, unable to return south. Mikkelsen proceeded to Victoria, BC, where he finally found an affordable vessel that showed promise, but only after extensive repairs and modifications. The former sealing schooner, built in 1879 and named the *Beatrice*, weighed 66 tons, and was 67 feet in length with a beam of 19.6 feet. The depth of the hold was 7.9 feet. Built in Japan of camphor wood, which is still used today in boat building, the boat's sides were covered with a protective layer of gumwood that extended three feet below and two feet above the water line. Mikkelsen later wrote that motorized ships in the early 1900s were "a comparatively new thing."[11] They could not afford to purchase an engine anyway. Sailing was not the fastest way north, but it was the best they could do.

Mikkelsen ordered many repairs and modifications, and the costs put a significant hole in the expedition budget—the vessel cost $2,600 and repairs and outfitting added about $2,000. The vessel could accommodate four men in the forward cabin and seven men aft.

Duchess of Bedford, Victoria, BC, 1906. *USGS Photographic Library, Leffingwell Collection*

One of the last details prior to departure was the critical, still unresolved issue of the flag under which the ship would sail. Mikkelsen was a Dane, so the Canadian ship could not be sailed under the British flag. It could not sail under the American flag unless ship modifications representing two-thirds of its value had been made in an American port. Mikkelsen sought advice from Secretary Keltie at the Royal Geographic Society on March 23, 1906, who suggested that Mikkelsen write the Duchess of Bedford. She agreed to be listed as formal owner of the boat so that it could fly the British flag. It was still necessary for Ottawa to pass an "order in council" before the flag was legal. The schooner's name was changed from *Beatrice* to *Duchess of Bedford.* If the tradition of changing the name of a vessel was bad luck, they were in for it.

The *Duchess of Bedford* was christened in a ceremony held May 1 at the Victoria Sealing Company wharf. The daughter of the Lieutenant Governor did the honors of cutting the line that smashed the champagne against the bow. Important Canadian dignitaries attended and photographs of the event revealed women in their finest dresses and men dressed in their best suits.

Preparations and Departure

At the time of the christening, Leffingwell and Mikkelsen were still hiring the sailing crew for the trip north. In addition to the scientific staff of

Duchess crew members, from left, Ernest Leffingwell, Captain Ejnar Mikkelsen, Dr. G. P. Howe, and Ejnar Ditlevsen. *USGS Photographic Library, Leffingwell Collection*

Leffingwell, Mikkelsen, Ditlevsen, and Howe, five additional members were added. Christopher Thuesen, age 27, had been with Leffingwell and Mikkelsen on the Baldwin-Ziegler Expedition. A Mr. Edwards was hired as mate, and "Storker" Storkerson, age 23, and J. Parker as sailors. The sailing crew also included an unidentified cook.

The schooner was loaded with scientific equipment, supplies, and provisions, some of which were shipped from Norway and Denmark. The scientific equipment included watches, a chronometer, sidereal watch, telescope, a plane table, alidade, sextant, and altazimuth.

Supplies included guns, rifles, ammunition, a phonograph, whaling harpoons and associated gear, articles for trading, skis and poles, wooden kayaks with canvas covers, dog harnesses, and senne grass, shipped from Bergen, Norway, for lining footwear. Expedition equipment included "Nansen sleds" (designed with wide wood runners coated with a thin silver alloy).

Provisions included pemmican, a staple of Arctic expeditions. Pemmican generally served as emergency rations and consisted of dried meat and fat. Horlick's in Michigan provided hundreds of pounds of malted milk. Other provisions obtained from Norway included twenty weeks of food in soldered tins for use when they traveled out over the pack ice. The contents of the tins were not reported, but likely emphasized pemmican. There were also tins of provisions for the dogs, and City Mills of Seattle reportedly donated 160 sacks of cornmeal feed for the pack. Leffingwell's father had purchased a large lemon ranch in southern California, so the family contributed cases of lemons.

Meals quickly became repetitious and included one or more of the following: potatoes, rice, onions, bacon, applesauce, prunes, cornbread, beans, and macaroni, supplemented by different types of codfish, shorebirds, and whatever game they hunted.

Sailing North at Last

The AAPE was the first Arctic expedition to sail from Victoria. Prior to departing, Mikkelsen made arrangements for return of the *Duchess of Bedford* from the Arctic back to Victoria. Word would be sent from Nome to a scientist in Glasgow, Scotland, to meet the vessel in Victoria. The scientist would re-provision the vessel and return north to retrieve Leffingwell and Mikkelsen from Siberia, Wrangel Island, or points

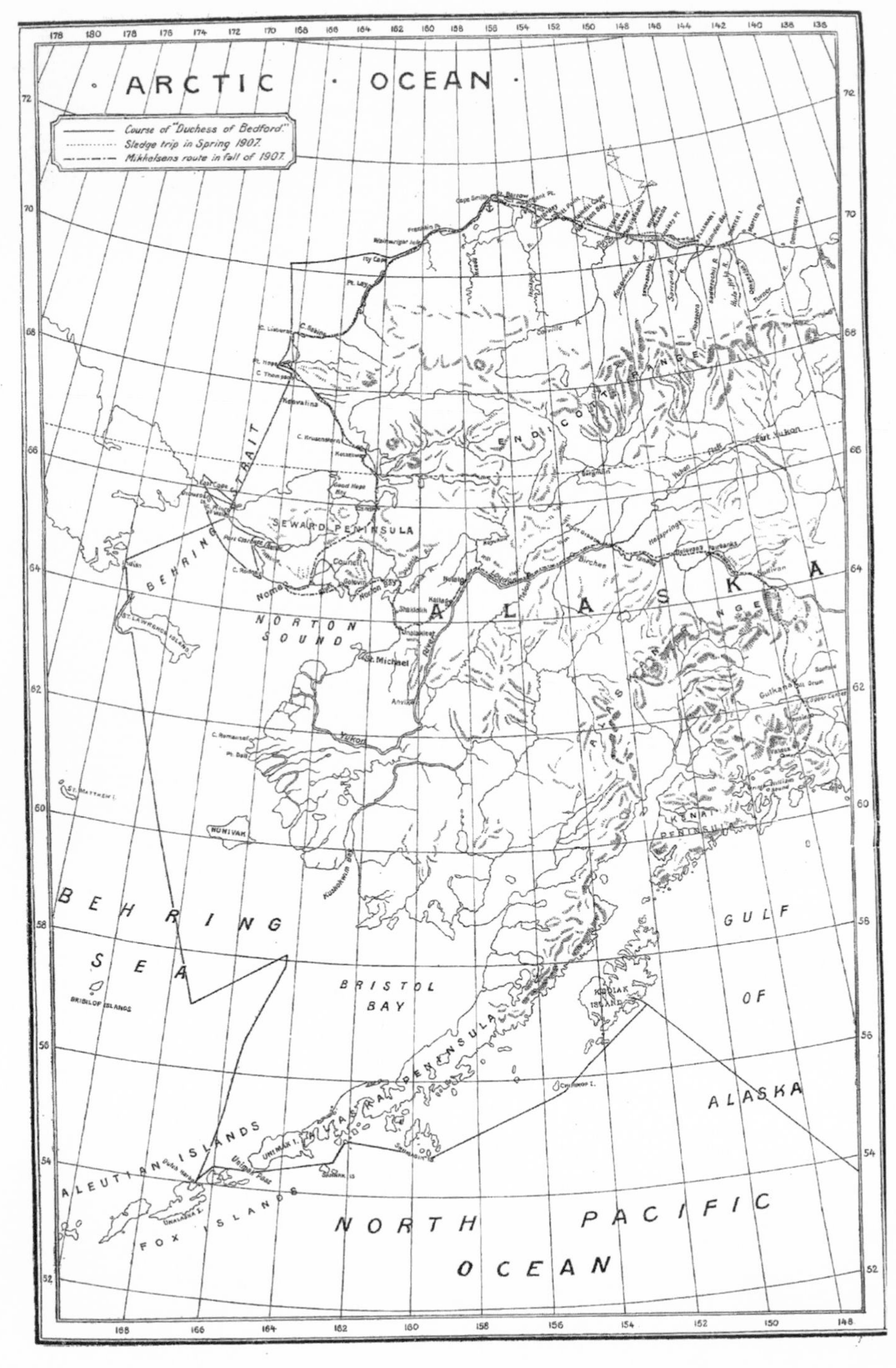

AAPE Route, 1906. *Mikkelsen, Leffingwell, and Howe, from the book* Conquering the Arctic Ice.

south. And if the Scottish scientist could not make it, he was to find someone who could. The RGS was to be consulted about all plans. Not surprisingly, the expedition would not unfold as planned.

Departure was originally scheduled for May 15, then May 17, then May 19, then delayed an additional day as they waited for vessel papers, additional provisions, and crew. Finally departing Victoria on Sunday, May 20, 1906, at 4:30 p.m., they immediately suffered an incident described by Leffingwell in his journal: "Tug ran us into dock & broke a stay but nothing else."[12] The party of eight had a short day of it and dropped anchor in nearby Esquimault at 7:00 p.m. Their first meal was malted milk while they waited for the cook to arrive from Seattle aboard a steamer. That evening, the cook arrived, making them a party of nine.

On Monday, May 21, they were still in Esquimault. Mikkelsen went to town to sign the final vessel papers. It rained all day but Leffingwell and Howe decided to climb overboard for a swim. They had dinner on a nearby vessel, the *Shearwater,* an Admiralty inspection ship of the Royal Navy.

The following morning, they were truly ready to leave for the north. Instead they "drifted around all day in front of Victoria" due to a lack of wind.[13] Leffingwell spent most of that Monday and Tuesday sorting and stowing gear. Quite a start for their expedition. A breeze finally came up about 9:00 p.m. Leffingwell's dry sense of humor resurfaced and he credited Howe, writing that Howe had raised the breeze "in return for a month's wages = 25 cents."[14]

Finally, on Wednesday May 23, they set off north under a light downwind breeze. Leffingwell kept busy stowing the last of the gear. By afternoon the breeze had changed to a headwind coming out of the west. Much to their surprise, the *Duchess of Bedford* started leaking enough that the crew was forced to pump the vessel regularly.

Leffingwell and Mikkelsen decided the leakage was not enough to warrant remaining in Victoria. They were anxious to be underway and realized their goal to reach the north coast of Alaska would be jeopardized by further delays.

They sighted whales in the Juan de Fuca Strait and observed albatross and petrels following the schooner. They "beat" or sailed against a gale with fifteen- to twenty-foot waves. They took down the mainsail and used the foresail and were still traveling at a rate of seven to nine knots, about eight to ten miles per hour. The gale lasted several days. Noting the continuing heavy swells, Leffingwell struggled with seasickness. He remained in his bunk during rough seas, sleeping ten to twelve hours a night, but given the motion even that did not seem enough.

Leffingwell estimated they had traveled 180 miles over two days, likely nautical miles; his journal did not distinguish between statute and nautical miles. The following week was a mix of weather and over five days they traveled a total distance of fifty miles and then endured another gale.

Kodiak Island

The Expedition reached Kodiak Island in a mist at 8:00 p.m. on June 5, so stayed offshore until the following morning. The following day Leffingwell wrote that their position was uncertain: "Beautiful scenery sim [similar] to around Tromso [Norway]. Heavily glaciated up to about 2500 ft. Mts. up to 4000 & snow at 2000."[15] They visited an Indian camp in an unidentified bay and purchased bear meat and salmon.

While Leffingwell organized gear, Mikkelsen, Dr. Howe, and Storkerson left on June 8 to hunt the grizzly bear promised to the Tring Zoological Museum. They returned about noon on June 11 without a bear. On June 14 the expedition left Kodiak Island headed for Unimak Pass (near Dutch Harbor, not quite halfway out the chain of Aleutian Islands). They promptly ran aground on a sand bar that morning at 10:00 a.m. "Two feet out of water low tide & 30° list."[16] By 4:00 p.m. they were off the bar, but the incoming tide kept them anchored for the night.

Unimak Pass was calm when they arrived, but they eventually determined that they had drifted fifteen miles backwards into choppy seas. Leffingwell noted, "Two other schooners near. All 3 boats with diff [different] winds."[17] By late afternoon a strong breeze rose, and the tides changed; they came through the pass and headed toward Dutch Harbor.

Dutch Harbor

The *Duchess of Bedford* finally arrived at Dutch Harbor late on June 21, and their stay was marred by the loss of one of the anchors. Leffingwell took observations to determine their geographic position, wrote letters, and visited the U.S. Coast and Geodetic Station at Unalaska.

During their stay Leffingwell mentioned for the first time in his journal that one of the sailors, Edwards, was suffering from malaria and that Ditlevsen, the artist and zoologist, was in acute pain from indigestion for the previous month. "I do hope Dit will not have to leave us. He is the finest man in the world & will help steady us when hard times come...All say we shall have hard luck in future."[18] It was a prophetic statement.

Before they left Dutch Harbor, the U.S. lighthouse tender *Heather* generously provided them with a new and heavier chain for their remaining anchor. As they sailed north through the Bering Sea over the next two weeks, they had a mix of sailing weather. "Five days of headwind. Rough. Can't stand up or sit down. Most disagreeable. Only 190 miles on course. All hands cursing. I still feel motion except when lieing [*sic*] down."[19] Days that followed alternated between calm and a pleasant breeze where they cruised at six knots per hour. They observed seals, sea lions, loons, whales, and caught codfish to supplement their meals.

The Fourth of July came and went with little fanfare. Howe saluted the day by firing a gun, and Mikkelsen thoughtfully gave Leffingwell a combined aneroid and compass as a gift.

They reached the Southwest Cape on St. Lawrence Island on July 5 and were met by five Yupiks. The expedition members traded calico (cotton flannel), tobacco, flour, sugar, and bread in exchange for seventy-five birds and twelve dogs, and in the evening visited the Yupik village.

As planned, the expedition then headed toward the Siberian coast, hoping to purchase additional dogs and a pony, but fog followed by wind when approaching Indian Point forced them to cancel any landing. They set course in the fog for Nome and Port Clarence on the Seward Peninsula to pick up freight and obtain more funds from Leffingwell's father.

In poor weather and visibility, they traveled north of Nome and Cape Prince of Wales. They were at the west end of the Seward Peninsula in choppy seas and strong current. Navigation became especially challenging as twenty fathoms (one fathom is six feet) became ten fathoms in about a half hour. None of the crew notified Mikkelsen. They suddenly sounded a dangerously shallow four and one-half fathoms. They carefully negotiated to deeper water. Leffingwell later wrote: "Were fairly sure we were on shoal N. of Cp. P of W, but map gives only 1 1/2 fath[oms]...Might have been on shoals near Port Clarence...Ejnar

[Mikkelsen] handled everything cooly & all kept their heads in spite of the fact that we were in a SE gale with only a foresail and [staysail] up. If we had grounded we should have gone to pieces quickly...Luckily for us the depth was greater than on map."[20] The lack of an engine was proving a hardship in navigation.

In the overnight fog, winds and currents carried them over 150 miles north of Nome, their destination. They were about 20 miles north of Diomede Island, and within three to four miles of the East Cape of Siberia. The next morning brought a calm, clear day with winds from the north for the first time in ten days. They passed the village on Little Diomede Island, at long last headed south to Nome, traveling at six and seven knots per hour. But the north wind did not last. They were southeast of King Island well on their way to Nome when the wind changed to the southeast. The barometer was falling. There was no way they could reach Nome.

Port Clarence and Teller

The expedition changed course once again, this time for Port Clarence, and anchored the same day (July 12) inside the spit, where they were greeted by a large herd of reindeer (caribou). They needed to travel overland more than sixty miles to reach Nome and conduct their business. It was now Friday, July 13, and they raised anchor and sailed across the bay for a pleasant visit with local missionaries, then rowed over to a whaling ship known as the *Wm. Bayliss*.

They shifted anchor to nearby Teller and transferred freight, part of a prepaid shipment that had been ordered for delivery to Victoria, from the U.S. revenue cutter *Corwin*. By the time the freight caught up with the expedition at Port Clarence, six boxes were missing, over thirty had been broken into, and many boxes had been tied with cord after falling apart. The freight bill was $295 and expedition members sent cables for desperately needed money to pay it. The bill was eventually paid with $190 of Leffingwell's father's money, and the captain of the *Corwin* agreed to take a note against the *Duchess of Bedford* for the balance.

While they waited for funds, Leffingwell made bread, took several time observations, developed and printed photographs, and hosted the captain of the whaling ship *Harvester* for supper. They also hosted the captain of the revenue cutter *Thetis,* who generously gave gifts of coiled rope, oars, venison, beer, and whiskey. Leffingwell confessed, "wish it was all over & that we could get the last link broken that connects us with places where money is used. We have everything else."[21] He was concerned that they would have to sell some of the provisions to pay their debts.

It soon became clear that Ditlevsen and Edwards would have to leave for health reasons. The cook and another crew member, Parker, also wanted to leave for Nome, where the gold rush was well underway. The cook, Parker, and Mikkelsen had an altercation on July 22, and a lieutenant from the *Thetis* was brought on board to settle things. He informed the cook and Parker that they could continue with the expedition or return south "in irons." They both preferred leaving in irons unless Mikkelsen would agree to let them go at Point Barrow. Mikkelsen promised he would, if he could find replacement crew members. The two decided to stay on board, and the *Duchess of Bedford* set sail, passing Cape Prince of Wales eleven hours later. They anchored south of Point Hope on July 24 and a group of white men and Inupiat visited the ship. The expedition acquired two large dogs and sixteen dozen guillemot eggs.

During this time, Leffingwell and Mikkelsen signed a new expedition agreement in which they agreed not to criticize each other in public, no doubt a result of their experiences with the Baldwin-Ziegler Expedition. Further, they agreed to appear together before the American Geographical Society and the Royal Geographical Society for resolution of serious disputes.

The expedition anchored at Cape Thompson along with the *Thetis* on July 25. Captain Hamlet of the *Thetis* determined that three of his men, Joe McAlister, Max Fiedler, and William Hickey, were willing to travel with the *Duchess of Bedford.* The new men were taken on board and papers were signed. The cook and Parker were put ashore at Point Hope the following morning. Leffingwell wrote that he and the others were relieved to see them leave.

The expedition continued its journey, encountering more north winds, swells, and ice, sometimes with little distance to show for their efforts. Leffingwell was pleased to see ice again. They saw over 200 walruses on ice floes, and at one point anchored near the village at Icy Cape. The expedition continued north and east, beating against the wind. By taking frequent soundings and traveling inside the grounded ice, hugging the shoreline, they managed to catch up to the other vessels near the Sea Horse Islands. On Sunday, August 19, they again sailed into the wind, finally reaching the village at Point Barrow, where they added Joe Carroll to their expedition as cook. They were disappointed to learn they would not realize their destination of Banks Island due to heavy pack ice that would not clear out until very late in the season, leaving a very narrow window for vessels traveling east. They also discovered that the maps and charts they carried were highly inaccurate, requiring extra diligence and careful attention to location, water depths, and ice conditions.

Barrow to Flaxman Island

As the barometer dropped, Leffingwell and Mikkelsen walked on the shore at Barrow, hoping for a change in the wind direction so they could proceed east.

What a shock it must have been for Leffingwell and the others to see Roald Amundsen arrive at Barrow aboard the *Gjoa*, having just completed the first successful traverse of the Northwest Passage. That search had been underway for hundreds of years, and Amundsen's traverse was a monumental accomplishment in the history of exploration. He would later cement his legacy as the first to reach the South Pole. In his typical straightforward manner and without drama or emotion, Leffingwell wrote, "Gjoa came in & finished NW passage. Ejnar, Dr. H & I were first on board. They left Herschel on 11th & saw Stefansson there."[22] After breakfast the following morning they were back on board the *Gjoa* for an exchange of gifts. While the AAPE had gifted a box of lemons to the *Gjoa*, the *Gjoa* gifted to them an ice saw, some dynamite, and guncotton.

The *Duchess of Bedford* attempted to continue its eastern progress but the currents were so strong they returned to anchor. They sailed east the next morning against the usual northeastern wind. The schooner increasingly showed signs of wear and tear. Leffingwell noted that it was leaking close to an inch per hour. The ship had hit grounded ice several times, once hard enough to throw Leffingwell off his feet. They were discouraged at their lack of progress and the relentless headwinds pushing them toward the coast. Their plans were in a constant state of flux. On August 28 they decided to return to Elson Bay, but found the shallow water, ice, and currents troublesome. They took up the anchor and quickly went aground. Leffingwell wrote:

> Had to blast floe with Amundsen's dynamite. Used fuse first trial & wrapped charge in paper. When I lit it & dropped it into water on top of ice foot, it started to float away instead of sink. Drifted toward ship but soon sank. Finally only cap went off. Next used electric detonators & broke floe nicely (50 ft away). Hauled off by noon & had lunch... Ran aground after bumping over one bar. Tried kedging [moving the ship using a line attached to a small anchor dropped at the intended destination], wouldn't work. But up all sail & after which got across bar, & had to drop sail quickly.[23]

With the *Thetis* anchored nearby, Mikkelsen released one of the crew, Joe McAlister, because of failing health. Early September brought snow and increasing ice. The captain of the *Belvedere* agreed to tow the *Duchess of Bedford* to Return Reef, fewer than twenty-five miles northwest of Prudhoe Bay. They started east on September 5, traveling slowly at three knots. When the *Duchess of Bedford* hit a piece of ice it made a loud noise, but the vessel did not appear damaged. In return for the *Belvedere*'s generosity, they gave the captain a pair of binoculars.

The *Duchess of Bedford* sailed east toward Barter Island and picked up a local family, Ned Arey, his Inupiaq wife, and two children. Leffingwell and Arey would become good friends and share trips and many conversations over the years following. The expedition passed Thetis Island, Cross Island, and encountered shallow waters, often only three fathoms.

Plans changed quickly and frequently. On September 8 they decided to winter on Barter Island so Leffingwell could explore the geology of the area and Mikkelsen could travel north over the ice looking for land. By the next evening they had decided instead upon Flaxman Island because it was closer and would provide the same opportunities. Another benefit was that two Inupiaq families also lived on Flaxman Island. The two families would become a meaningful and important part of their lives helping to ensure the survival of expedition members. They would become friends and trading partners, helping each other out in times of need.

Fog and shallow waters impeded progress. On September 11 Leffingwell wrote, "Cleared off later for first time since Aug 2 at Icy Cape. Could see mountains 40-miles away. Beautiful. Fog again after supper."[24]

Arey and his family left the schooner near Flaxman Island to continue their journey east. When the *Duchess of Bedford* anchored, Storkerson took the small boat to take soundings of the route to Flaxman. With Storkerson busy, Leffingwell and Mikkelsen traveled to the mainland, built a fire, and discussed the idea of Leffingwell accompanying Mikkelsen to the north in the spring. That evening they dined on what Leffingwell referred to as reindeer meat, likely caribou. While Mikkelsen and Howe played chess, Leffingwell and Storkerson played checkers. There was heavy fog again September 12. Storkerson, with some of the crew, traveled to the mainland looking for a possible winter shelter. They had no luck, so Storkerson and the others continued taking soundings toward Flaxman Island. On September 14 the first wind out of the west since the end of July pushed the *Duchess of Bedford* gingerly three miles closer to Flaxman Island. Leffingwell noted the beautiful sunset, a rare sight due to the frequent fog.

Leffingwell at Flaxman Island, 1906. *USGS Photographic Library, Leffingwell Collection*

They grounded again on September 16, but finally, at 10:00 a.m. on September 17, they anchored in the lagoon, on the south side of Flaxman Island, in eight feet, five inches of water. Leffingwell wrote that the *Duchess of Bedford* drew seven feet, ten inches, which they had achieved by hanging the small boat and chain under the bow, thereby raising the stern five inches. Every inch counted in the shallows and grounding was always a factor. Mikkelsen promptly set a pole for gauging tides, which he monitored hourly. Leffingwell explored the island to the west, making mental note of dandelions and arctic poppies. He was also pleased to discover glacial boulders. He collected some specimens and returned to the ship.

After nearly four long months en route to the Arctic, they had finally arrived to establish their winter quarters on the south side and western end of Flaxman Island. It was not their original destination, but given the weather and ice conditions, they would not be sailing any farther east. Leffingwell and Mikkelsen had proved that they were problem solvers and strongly motivated to succeed in their goals.

Their intended goal of determining if there was land north of Alaska remained a viable option, even if they were not traveling from the west coast of Banks Island as originally planned. And before their search for the edge of the continental shelf ended, Leffingwell knew that being out on the pack ice was something he never wanted to experience again.

Flaxman Island topographic map, Alaska. *USGS*

CHAPTER 4

Winter Quarters at Flaxman Island and the Search for New Lands

Expedition members lived on the ship in the Flaxman Island lagoon at the island's southern end during the winter of 1906. The lagoon protected them from Beaufort Sea ice movement onto the north shore.

Having left five original members at Point Clarence and Point Hope, and with the *Thetis* providing several new members at Point Hope, the expedition now consisted of Leffingwell, Mikkelsen, Howe, Thuesen, Storkerson, McAlister, Fiedler, Hickey, and Carroll. Their arrival on September 17 was celebrated and accompanied by an aurora borealis display that evening. The ice had not yet frozen in; there was still open water and they could boat between the ship and shore. The crew started unloading some of the goods from the leaking ship.

On September 21, Leffingwell started building his observatory ashore. "Calm & clear. Built pier for observatory. Log 10 in diam [diameter] & 6 ft long, set 2 ft in ground on frozen soil."[1] At noon he took observations for latitude.

The two Inupiaq families that lived on the eastern end of the Island, Shugvichiak and his wife Toklumena, and Uxra and his wife Toolik, greeted the newcomers. Shugvichiak quickly became a good friend whom Leffingwell called "Sagi."[2] With the help of the Inupiat, the next major task of the expedition was to supplement their food supply. Two days later, Mikkelsen and Howe left on a hunting trip with Uxra and Toolik. They took three dogs and planned to be gone until the freeze up occurred. Meanwhile, Leffingwell dug a well two feet deep in ground ice for fresh water. He spent his leisure time reading *Barren Grounds of Canada* by Warburton Pike, and began working on a geology article about Flaxman Island and glaciation.

Leffingwell's article, published in 1908, described the setting: "Flaxman Island is about three miles long and half a mile wide, running nearly parallel to the mainland. Its surface is a tundra plain about twenty feet above the sea. Ponds are scattered over the surface, and large crystalline bounders are frequently met with lying half buried in the soil. Immediately over the beach the plain ends in a steep mud cliff which is broken by frequent gullies.[3]

Leffingwell presumed that ground-ice underlay the whole island to an unknown depth. During the summer only the upper foot or two would thaw.

Triangulation station erected over astronomical observatory pier at Flaxman Island. *USGS Photographic Library, Leffingwell Collection*

Weather dictated their schedule, and gale force winds from the east occurred every few days. Although the *Duchess of Bedford* had run aground, and they had placed two anchors, one of the gales moved the vessel about 20 feet closer to shore. The ship may have been aground, but it bounced hard when the gales hit. Leffingwell estimated that the winds from the various gales were blowing 40 to 50 miles per hour. He wrote of one incident where two of the men had gone ashore to take observations. When the gale came up, instead of taking 20 minutes to travel about a mile, it took them 20 minutes to travel fewer than 100 yards back to the ship.

Shugvichiak brought the expedition members 30 pounds of "deer" (caribou) meat, and Leffingwell gave him tea, malted milk, sugar, and other items in return. Shugvichiak returned with a seal two days later, and took Leffingwell's measurements for warm winter clothing. Leffingwell provided him with some reindeer (caribou) skins and Shugvichiak returned several days later with a coat, pants, and gloves sewn by Toklumena. Leffingwell judged that the reindeer fur coat weighed five and one-half pounds and the pants two and one-half pounds.

During October, Leffingwell finished writing his article, developed photographic plates, taught Storkerson navigation, and worked on dog harnesses. He started training the dogs. Mikkelsen and Howe returned on October 8 with only one dog. One dog had run away and the other had died from a disease known as "moly coly," later determined to be rabies. The hunting efforts by Mikkelsen and Howe had yielded one caribou.

Ice conditions made travel between the ship and shore increasingly more difficult. It was too thin to walk on, but becoming more difficult to row through. They awoke on October 11 to four inches of snow, and water that had risen enough to float the ship. By October 14 the grounded ice still would not hold Leffingwell's weight, so he stayed on board. Uxra and his family braved the ice conditions and arrived at the ship with a sledge and three dogs. The following day, Shugvichiak's wife Toklumena brought some seal meat (Leffingwell wrote "oogruk meat") and additional clothing for Leffingwell: a snow coat and his fur coat. He kept his feet warm by wearing rubber boots with one pair of socks and lined with senne grass.

As he explored the island, Leffingwell described the ice as "coarsely granulated" with "minute air bubbles" and inferred that the source was either glacial ice or snow. Leffingwell observed crystalline boulders embedded in ice, ponds, and mud cliffs, examined other rocks consisting of conglomerates, quartzite, limestone, and granitic composition, and found striations on many, indicating glacial deposition. He found that the base of the ice was not visible in any location on the island. Leffingwell concluded that Flaxman Island was a barrier reef created by waves, had been covered by a glacier at one time, and that the source was either the Sadlerochit Mountains to the south or the headwaters of the Canning River, which he referred to in his article as the Kugura. Modern-day geologic maps note that Flaxman Island is composed of glacial drift.

Temperatures had dropped to -10° and Leffingwell was hoping to leave for Herschel Island on October 17. He intended to meet up with Stefansson, drop off mail, and secure additional supplies. Leffingwell wrote, "Decided to give up surveying as we go along, time is short, mail leaves Herschel Id on Nov 1st and we may be delayed by storms. Besides light is poor & short."[4] Mail went out from Herschel Island with the North-West Mounted Police to Dawson before continuing south. Sending and receiving mail took months. Mail from northeastern Alaska traveled via several routes: by river, overland, and via the whaling ships and revenue cutters.

Fall 1906 Trip to Herschel Island

The distance from Flaxman Island to Herschel Island is over 170 miles. From there Leffingwell and Mikkelsen hoped to send news of the expedition to their supporters and family. Leffingwell, accompanied by Howe, Storkerson, and Thuesen, departed on October 17, but their hopes of reaching Herschel Island by November 1 were soon diminished by weather. Bitter temperatures were holding. Leffingwell described ice conditions varying from sticky to fresh, to thin, to non-existent, to soft and slushy, which affected how the sledges traveled, how fast the dogs pulled, and the routes they took. When the ice conditions were not safe, the team settled into one location for one or more layover days.

Leffingwell experimented with clothing, trying to get the right mix of warm but not too warm. He wore fur clothing from a variety of animals, including seal, squirrel, and caribou. Leffingwell and the others quickly adopted the local Inupiaq clothing, as it was far superior to anything they brought for cold weather and travel. During the Baldwin-Ziegler expedition, they had all worn wool, which was inferior to fur for warmth. Clothing on this expedition was made of caribou or sheep skins, and included shirts, atteges (hooded pullover, knee-length, fur-lined overcoats), pants, mittens, and socks. Boots were made of seal skin, and Leffingwell estimated the total weight of clothing and boots at 10 to 11 pounds. The fur was always worn to the inside for warmth. If a second layer was worn, the fur was to the outside.

Leffingwell observed the aurora borealis and moonlight. As they traveled, he sketched a map of the mountains to the south and the shoreline for future use. They met up with Ned Arey and his son Gallagher at Simpson Cove, south of Collinson Point, on October 22. After more bad weather and a trip to retrieve a cache of meat, Arey, Howe, and Thuesen began their return to Flaxman Island on October 29. Leffingwell, along with Storkerson, stayed another day before continuing east towards Herschel Island.

The two reached Arey's cabin at 11:00 a.m. and promptly lit the stove to dry out their sleeping bags and clothing. They had crossed the river channel five times and were at times traveling in snow eighteen inches deep. Arey's cabin was about sixty miles from Flaxman. They had clearly missed the departure of the mail from Herschel Island. It would be months before the mail went south again.

As their journey east continued, the distances they traveled each day varied significantly, depending upon wind, snow, and ice conditions. If the weather was good when they made camp, the tent was erected in about thirty minutes. Temperatures inside the tent depended upon whether they were camped on land, on the grounded floes, or in a location protected from weather. It was -28° on November 6 and the sun set at 1:45 p.m. Leffingwell wrote: "Clear & quiet & coldest yet. Used fur coat for first time. Did not sweat today. going very heavy...load dragging in snow. Both hauling. Dogs quickly tire & we also. All one man could do to start sled. Went 4 hours at first 2.5 miles last 1 1/2. Felt used up. Am not hungry but have no bottom."[5]

The conditions along the north coast of Alaska were very different from what Leffingwell had endured in Franz Josef Land. On November 8, Leffingwell wrote, "Burned blubber lamp all night & kept fire. very tiresome. How did Nansen stand it a whole year?" He was discouraged and frustrated. "14 days grub left on sled must do 10 miles calc [calculate] we are 70 miles from H Island. If going were good could do it in 3 days. Both feel blue over way things are going. Instead of having a pleasure trip it is getting monotonous & poor fun. Never looked for soft snow in Nov. On FJ Ld [Franz Josef Land] snow was packed in last of Sep."[6]

When a gale hit on November 11, they hoped the wind would pack the snow, which it did. As they traveled toward Herschel Island they picked their way through the offshore ice as best they could, negotiating pressure ridges formed when pack ice crashes into the shore ice and pushes up a ridge of ice. Sometimes the sledge overturned.

On November 18 Leffingwell calculated that the distance to the whaling ship was 15 miles. He had also observed open water near Herschel Island and was cautiously optimistic about making their destination the following day. That evening, however, a gale came up and Leffingwell lamented his campsite. As the wind increased to 45 miles per hour, the temperature fluctuated between -12° and -23°, without the wind chill factored in. The gale and snow continued for several days. So they hunkered down and waited.

Herschel Island at Last

Leffingwell and Storkerson arrived on board the whaler *Narwhal* anchored at Herschel Island on Wednesday, November 21, after five weeks of travel. Captain Leavitt had the cook prepare a huge meal. He gave them a change of clothes to wear while theirs were mended and offered them cigars. The following day Leffingwell read in the captain's cabin before going ashore with Storkerson and another man amidst a gale. He later described it as "the worst storm but one I have seen in winter. Other 75 mile gale at Franz J Ld [Land]."[7]

Leffingwell spent much of the following day at the police headquarters, where he learned Stefansson was wintering on the Mackenzie River with A. H. Harrison, a British explorer, who was also in the Arctic looking for new land and planned to travel to Banks Island.[8] In a letter to Mikkelsen, the Royal Geographic Society had suggested that they combine their efforts, but with the *Duchess of Bedford* grounded in ice for the winter south of Flaxman Island, circumstances prevented the possibility.

On Saturday November 24, Leffingwell wrote: "Cold & still W wind with snow drifting up to knees, might have gone west didn't relish idea, worked on repairs, etc. Cleared off pm quiet night & beautiful moonlight went to native Hula Hula dance. About 40 people in house 15x15 & no air."[9]

After a week at Herschel Island, Leffingwell and Storkerson started their return journey on Wednesday, November 28. Captain Leavitt and his first mate provided them with two additional dogs, a carpenter who repaired their stove, some tobacco, and some feed for the dogs. Leffingwell was concerned that they had overstayed their welcome so he presented Captain Leavitt with a gift, his Bausch and Lomb twelve-power binoculars.

On the first day, weather and snow conditions were good and they covered an amazing thirty-three miles. Later, they were fortunate to travel four miles in seven hours (at -35°). They reached the lagoon at Demarcation Point and camped for several days as they waited out the latest storm. The cold winds froze their noses and chins.

On December 10 Leffingwell and Storkerson reached Barter Island after traveling two miles in eight and a quarter hours in -34° weather. By the following mid-day, they had reached Arey's cabin. They spent four days in the cabin, drying and mending gear, and resting the dogs and themselves. On December 15 they rounded Brownlow Point about noon and reached the *Duchess of Bedford* by 1:40 p.m. Their journey to Herschel Island had taken them eight and a half weeks, much longer than Leffingwell had ever imagined—a powerful experience that tested all of their skills.

Back at Flaxman Island

Back aboard the *Duchess of Bedford*, Leffingwell spent the first week relaxing and reading and occasionally working on the altazimuth. He decided to shave off his moustache and beard after experiencing the cold and ice around his mouth during the trip.

Leffingwell carefully monitored conditions and supplies and calculated that they were burning fewer than 80 pounds of coal each day; half in the galley and half in the cabins. He noted that there were six hours of twilight during the day and temperatures were -20°C to -30°C.

Christmas came and went with dinner of caribou, goose, plum pudding, pie, and fruitcake. The crew enjoyed a couple of quarts of whiskey and their celebration was jovial. On Christmas Day several Inupiaq families dropped by for lunch and an exchange of gifts. The afternoon concluded with Leffingwell playing the phonograph.

Mikkelsen and one of the crew, Fiedler, left the following day for Arey's cabin. Leffingwell returned to taking astronomical observations when weather cooperated, assisted by Howe. Leffingwell's time accuracy was impressive, within .8 seconds, and the latitude accuracy within 1.7 seconds of a degree.[10]

In the evenings, Leffingwell read about Elisha Kent Kane's Arctic trip in search of Sir John Franklin and compared some of their experiences with his. He wrote a self-deprecating description: "judging by my own abilities which I place as high as those of a scurvy ridden crew of sailors."[11]

Over several days, Leffingwell and Howe visited Shugvichiak's house. The women were making atteges, breeches, fur stockings, and kamiks (moccasins from animal skins) for winter weather and travel. Leffingwell enjoyed the trips and learning the Inupiaq language.

It was now 1907, and they had been away from home for over six months. By January 2, Leffingwell wrote that his "lazy spell" had passed. Bad weather consistently interfered with taking observations, but he went for walks, cleared snow off the deck, tested clothing at different temperatures and winds, and drew plans for a tent. He spent leisure time reading *Don Quixote*.

By Friday, January 18, Leffingwell was aggravated by the weather and his inability to work on occultations. He had not been able to make any observations since the first of January, so he set up some practice ones. Honing the skills would be critical for him in determining their latitude and longitude on the pack ice and during other travels.

After days of heavy blowing winds, Leffingwell observed the sun for a brief time on January 21. His concern about Mikkelsen and Fiedler was increasing and he considered sending a relief effort. Fortunately, they returned early the next morning with Gallagher Arey and another Inupiaq, but with a harrowing tale. When the gale arrived on January

14, their tent blew down, was torn, and soon filled with snow. They attempted to locate Arey's cabin, but visibility was poor; they slept in their bags out in the open. Leffingwell wrote, "Before they hit Ned's they were nearly gone. Ejnar's heel frozen & both wrists. Close shave."[12]

The Search for Land and the Continental Shelf

The purpose of the expedition was to determine the edge of the continental shelf and to search for land north of Alaska. If they found the continental shelf close to the Alaskan coastline, they were unlikely to find land north of it. As they traveled north on the pack ice, the edge of the continental shelf would be determined by taking soundings—ocean depth measurements. They used sounding apparatus that Mikkelsen designed in Copenhagen. He described it: "The line, an eighteen-stranded copper wire about 1 1/2 m.m. thick, with a tension of 40lbs., passed over an indicator on the end of an arm attached to the machine, so that any time we could know how much wire we had out. Our wire was 620 metres long. The sounding machine was lashed on to the hindepart of my sledge and was always ready for use."[13]

They planned to spend two months on the pack ice, after which Leffingwell would head to the mountains for June and July. Then, when the pack ice moved off shore, likely in August, the group would travel east to Herschel Island and continue to Banks Island to spend the winter as originally planned.

Preparations, especially obtaining more dogs, began in earnest for the sledge trip northward over the pack ice. Gallagher Arey interpreted for the crew as they traded guns, shells, flour, tea, and a watch for sled dogs. Leffingwell wrote, "Sagi [Shugvichiak] said he wanted to help us and insisted on giving us the best of his 3 dogs and asked nothing in return. Brought him back to supper. Gave him my 405 and 300 shells."[14] Leffingwell and Mikkelsen also "borrowed" a large dog in exchange for sacks of flour and two small dogs.

Their preparations were hindered when four feet of water were discovered in the ship's hold. Leffingwell suspected that the caulking had been pulled out by temperature contraction. He wrote: "We have no copper sheathing [to protect the underwater hull]. Rigging is extremely taut from contraction, hums at fairly high key in wind. May have sprung sides. We are on bottom so can't sink, but may turn over when ice breaks up...Crew got all food on deck before breakfast. Sent them bottle of whiskey & now they are hustling the coal out, singing & enjoying it immensely. Shall cache everything on shore. Cabin floor above water so we are ok aft. As long as she stays right side up."[15] Some of the Inupiaq families came to help sledge provisions from the ship to the Island and when finished, all food and half of their coal supply was ashore.

Leffingwell planned to take a sledge to Arey's with provisions for the upcoming trip. When Leffingwell, Howe, Gallagher Arey, and a friend of his loaded sledges on January 28, they determined that the three sledges totaled 2,400 pounds. The sledges would be pulled by fourteen dogs and the men would help haul as necessary. Their departure was delayed until the weather improved on January 30 and the party returned to Flaxman Island on February 10. Their experiences would be much different once on the pack ice.

Some of the sledges had wood runners and others metal runners, which meant that they responded differently to snow and ice conditions. Leffingwell found that wood runners worked well on dry snow but not on ice with a high salt content. Metal runners worked better on the salty ice.

Leffingwell took azimuth compass readings and determined the declination of the compass to be east 34°. It is necessary to add the declination to the compass reading if the declination is east in order to determine the location of true north. Leffingwell also compared timing on the four watches that would be used and found three of the four to be very consistent. Watch accuracy was equally important for determining their physical location.

During pre-trip scouting, Mikkelsen and Howe walked about six to seven miles out on the pack ice and reported difficult travel. The pressure ridges were tall with broken ice; the plan was to get through them on to the floe ice, which was relatively smooth. Mikkelsen and Leffingwell also scouted to observe ice movement and conditions.

Leffingwell was much more interested in exploring the mountains and landscape to the south. On February 25, Leffingwell he wrote: "If we come back [from the pack ice] without some important discoveries, I want to stay another year or so. At the N foot of a very prominent mt 70 miles inland—about 7000 ft. (estimated) there lies a large lake.

Pressure ridge, Beaufort Sea, 1907. *USGS Photographic Library, Leffingwell Collection.*

Reported to be a long day's sledge journey over it, so it must be about 25 miles long. Abounds in large fish nearly length of a man. Ugaruk & family will be there all summer, shooting mt. sheep. hope to see him there. No white man over there."[16]

On February 27, Shugvichiak came by the house and said that he would accompany them for a distance to "test the ice." Leffingwell and the others were relieved to have his guidance and wisdom. It was decided that Storkerson, Fiedler, and an Inupiaq youth, Okalishuk, would help them through the pressure ridges for the first day or two.[17]

Pack Ice Trip

At long last, it appeared that they would be able to start out on the expedition mission. Leffingwell, Mikkelsen, Howe, and Storkerson began their journey to the pack ice on March 2. Shugvichiak and Okalishuk and Hickey, another crew member, traveled with them for a short time.[18] Shugvichiak assisted in determining the best route. The party managed three sledges and 13 dogs; each sledge was loaded with an approximated 733 pounds. They took soundings to monitor the depth of the water as they traveled. When they arrived at open water, soundings were easy; at other times they had to cut a hole in ice that was sometimes three to four feet thick.

Based on earlier observations Leffingwell and Mikkelsen believed that the ice would be worst along the shore. This turned out not to be the case. On two occasions early in their journey, they crossed ice so thin

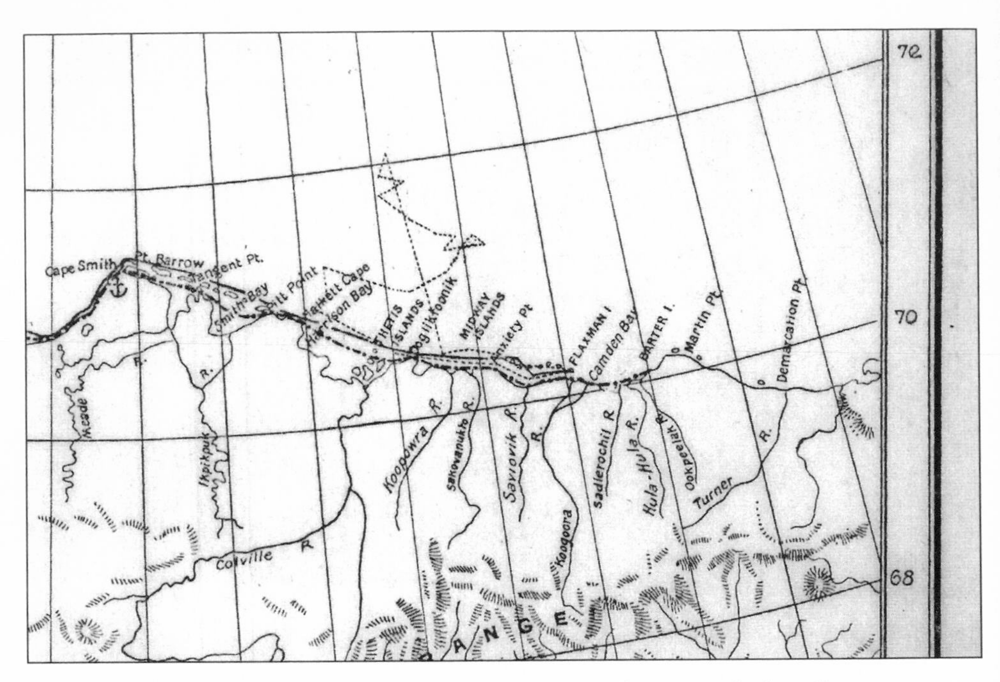

Pack ice trip route, 1907. *Mikkelsen, Leffingwell, Howe,* Conquering the Arctic Ice

it buckled under their weight. And they came upon lanes of open water every mile or so. The party was able to bridge an open water lane twelve feet wide by placing blocks of ice to make a path across the open water.

Leffingwell referred to different types of ice they found. One he identified as "paleocrystic" or "paleo," an old ice floe with hummocks, rounded domes of ice up to thirty feet high. (Hummocks of silt and clay topped with dense vegetation are also found on land.) Another he identified and described as a beautiful azure or "blue ice," which was young ice. A third was "salt ice" which was a young ice, recently formed. Another was "pan ice," recently formed as circular ice pads up to ten feet across. The old floes were much safer and stable when compared to the young ice.

They had traveled only one mile in five and a half hours when they reached a one-half mile wide stretch of thin ice on March 4. They camped for the night and determined their latitude to be 70°18'5"and their longitude an approximate 146°. But on the fourth day, they

Sounding through a hole in ice, Beaufort Sea, 1907. *USGS Photographic Library, Leffingwell Collection*

stumbled up to what Leffingwell referred to as the "Chinese Wall," a reference to the Great Wall of China. The block of solid ice ranged from four to twelve feet high. The party was at a critical juncture; they could traverse the daunting pressure ridges with monumental effort, or return to the ship. The group decided to push north.

Leffingwell described the terrain as a "chaos of ice block." In one hour they had cut a road 75 yards long. After lunch, the dogs and four men hauled each sledge, and traveled an additional 100 to 150 yards. After five long hours, using shovels and picks, they had worked their way through the wall. Leffingwell wrote: "In the pack about Franz J Land I saw nothing approaching that ice, nor coming up here last summer, nor ever read anything like it (except for a few hundred yds.) The average elevation of the floe was about 6–7 ft above the water, with a relief of about 5 ft along the road we cut."[19]

Jagged ice blocks extended north as far as they could see, a distance Leffingwell calculated at three miles. They figured it would take a week to traverse the area, causing the support party to be out much longer than planned, and the sledges, already breaking down, would be destroyed.

Dog team in difficult terrain near Beaufort Sea, 1907. *USGS Photographic Library, Leffingwell Collection*

Rough ice and soft snow, Beaufort Sea, 1907. *USGS Photographic Library, Leffingwell Collection*

Brecciated ice in the Beaufort Sea, 1907. *USGS Photographic Library, Leffingwell Collection*

On March 7 they decided to return south to the ship. The ice conditions were so rough that Leffingwell's sledge could not be turned around; it had to be hauled out backwards. During their retreat they encountered a new lane of open water, twelve feet across, and built another bridge of ice blocks across the open lane.

The party covered the seven miles to the ship in one day. They had been out for five days and were frustrated by the experience. Howe expressed his preference to not travel with them on the pack ice again, citing friction with Leffingwell. Leffingwell accepted his resignation. Leffingwell was clearly uncomfortable on the pack ice and it may have been reflected in his behavior. Following their return, a gale came from the west packing winds of 50 miles per hour, then one came from the east with winds of 25 miles per hour. Leffingwell was grateful that they were on the ship and not the ice. He used the time to read books, develop two dozen photographic plates, and make 40 prints.

Leffingwell and Mikkelsen now made a third attempt to determine the edge of the continental shelf and whether there was land north of Alaska. They would succeed in reaching the edge of the continental shelf, but it would be an arduous journey. In anticipation of the travel conditions they knew were ahead, they reduced the weight of each sledge to 600 pounds. They were up at 3:00 a.m. on March 13. The temperature was -15° and when they departed for Cross Island the wind started blowing. They eventually left the loaded sledges behind and returned to the ship to wait out the gale. It blew thirty miles per hour all day and continued to blow for the next three days, leaving a snowdrift extending half the distance to the crow's nest. Leffingwell spent the days reading and made plans for the fieldwork he hoped to accomplish in summer 1907, exploring the mountains to the south. It made sense, as his geology field work in glaciology had been completed in the mountains of Colorado, Wyoming, and Washington State.

When the weather slackened on March 17, they set out again in -28°C. Their provisions included food for sixty-five days, to be supplemented by seal and bear. The barometer dropped, the wind came up, and snow fell. They camped and remained three nights in the same location until the weather improved.

On the morning of March 21 they spent two and one-half hours digging out the sledges and the tent before starting northwest, reaching Cross Island in mid-afternoon. They established camp and cached a week of food for their return. They stayed another three nights as drifting snow and winds made visibility poor and travel conditions miserable.

Finally, on March 28, Leffingwell, Mikkelsen, and Storkerson spent their first day off land and on the pack ice. In keeping with his meticulous nature and training as a scientist, Leffingwell recorded the day of the week, date, camp number, distance traveled and time, temperature, winds, weather, departure time each morning, and the time of day that they stopped to camp. Even more detailed was his description of the conditions they endured each day. An example:

7:40-8	Cross heavy moving ridge. axe
8-9:00	Young ice, leads, fill in etc axe 1/2n mi.
9-10:30	good older ice, going etc 1.5n mi.
10:30-12	Blocked moving ridge
12-4:30	Young ice low ridges @1.5, wide floes 7n mi.[20]

They carried four watches that Leffingwell carefully monitored and compared for accuracy. If the sun was shining at noon, they took observations for latitude using a sextant. If it was not, they waited for an hour or gave up. At times they calculated their location by "dead reckoning," which required the use of a known or fixed location of a feature, the distance they had traveled and the rate of travel or speed. At the time, they were not concerned about determining longitude because they had not yet realized that the pack ice was moving, no matter how stationary it appeared. Longitude was more difficult to determine because it was necessary to know their latitude and know exactly when the sun was at its highest in the sky.

Leffingwell's notebook included photography tables that detailed a formula for determining aperture and an approach to determining shutter speeds. He correlated different shutter speeds to factors that included cloud cover, open seas, distant snowy mountains, and distant landscapes. And once on the journey, he maintained a log where he recorded subject, aperture, and shutter speed for each photo.[21]

Travel on the pack ice was a dangerous endeavor. During their first few days, Leffingwell, Mikkelsen, and Storkerson negotiated young, thin ice and open water interspersed with older and more stable ice floes. As they traveled they observed bear tracks, fox tracks, and a seal. They took a sounding on March 29 that revealed a depth of fifteen fathoms. Leffingwell was struck with snow blindness—a painful, temporary loss of sight resulting from overexposure to sunlight—so he started wearing protective "snow glasses."

The ice was constantly moving, floe against floe. When the ice collided, the pressure would force the ice up and create barriers of "screw" ice that were difficult or impossible to cross. When the heavy sledges occasionally tipped over, they had to be unloaded before they could right them, and reloaded before continuing.

Open lanes of water had to be negotiated; sometimes they built bridges using ice blocks, or they waited until one floe bumped up against another. When the gaps closed they were able to cross. Not only did the open lanes of water slow them, but so did soft snow with mounds or hummocks up to 20 feet high.

On March 30, one of the floes suddenly started cracking all around them. Progress halted for one and one-half hours as they waited for the section to close. Leffingwell monitored the movement of the floe—one foot every eight seconds. The trio camped on an old floe at 4:30 p.m., and Leffingwell felt that their understanding of ice movement was improving. They calculated their latitude at 71°00' north and the longitude at 149° west, and temperatures ranged from -35°C to -25°C.

The sleds often became stuck in the uneven ice, and the men did not fare much better. There was no stable footing and every several steps they broke through between blocks of ice along cracks. Sometimes they struggled through deep snow drifts up to their hips. Leffingwell was snow blind in his left eye, his vision blurry and painful. He wrote the following day that his eyes were improving. But the traveling conditions were not always miserable. Occasionally, the older "good" ice and flat areas provided relief. Considering the challenges, the men were all in good physical condition, although Leffingwell mentioned that his shoulders were sore from the strap used to haul the sledge. When the wind picked up again, they spent a couple of days at the same camp, dried and repaired gear, and relaxed. Leffingwell read *Hamlet* and wrote, "Quite a delightful time in camp in spite of penetrating E wind outside."[22]

They were 43 miles from land, and had been taking soundings that indicated depths at 30 meters, 44 meters, and then 86 meters. They were surprised when a few miles north, they found the edge of the continental shelf. Their calculated position was latitude 71°22' north.

They traveled eight miles on April 8, good progress, but when Mikkelsen took observations, Leffingwell determined they were at a longitude of 151°, farther west than they thought. They attributed it to "drift." But when they compared the opposite sides of cracks, they matched up, which indicated that there was not any drift occurring. It did not make sense.

The Continental Shelf

On April 10, their fourteenth day on the ice, they figured that they had traveled 86 miles. They took a sounding, and recorded "no bottom" at 620 meters. Their estimated latitude was 72°03' north and longitude at 150° west. Leffingwell, Mikkelsen, and Storkerson had reached the edge of the continental shelf, much closer to the coast of Alaska than they had envisioned it would be. They were surprised and questioned the results of their efforts. In a report to the Royal Geographical Society, Mikkelsen later wrote:

> We had not expected to find so deep water so close to shore, and had not used every opportunity we had to get soundings, but this deep sounding would probably indicate that we were off the Continental Shelf. We thought that there was a possibility of its being due to a submarine valley from the Colville River...Made wise by our disappointment in not getting the grade of the Continental Shelf, if we had passed it, and if not, to get the other side of the submarine valley, we took several soundings, but with no better luck.[23]

Using a sextant, Mikkelsen took a reading for longitude while Leffingwell calculated the results on April 9. They were surprised to find that that their longitudinal location was west of their starting point by twenty miles, not east as their calculations using dead reckoning had indicated. The breakup of the pack ice had occurred earlier than usual in 1907 and it was drifting. Mikkelsen explained: "We had been too firm believers in the general accepted idea of a practical immoveable pack, to pay much attention to our longitude but it was now evident, that we had drifted to the north as well as to the west, and that accounted for the supposed underestimation of the daily marches."[24]

Their prior assumptions about pack ice were wrong and the consequences were daunting and dangerous. There was a real possibility that Leffingwell, Mikkelsen, and Storkerson might not be able to return to the north coast of Alaska. They continually encountered open lanes of water as far as they could see the following day, April 10. Leffingwell and Mikkelsen concluded that they must be north of the continental shelf. They had traveled 32 miles north of where they had determined the edge of the continental shelf. The party decided to return south as travel was both challenging and life threatening.

They had accomplished their goal of determining the edge of the continental shelf, which in turn suggested it was highly unlikely that there was land north of Alaska. They planned to return the following year to explore the edge further, but for now it appeared that they had at long last answered the controversial question debated by explorers, geographers, and scientists—land north of Alaska was a myth. In spite of their findings, within several years Stefansson would obtain funding for the Canadian Arctic Expedition to once again look for evidence of land north of Alaska.

Return to Land

Returning to the Alaskan coastline and firm ground would prove as challenging as the first half of the trip. The sledges were falling apart. Wood chipped off in the ice rubble as they traveled. "Decided that we can make greater dist [distance] by saving sleds, as they are going faster than food" wrote Leffingwell.[25] Their efforts were slowed when several of the dogs died from "moly coly," or rabies. And during the pack ice trip they had lost a pick axe and the lead at the end of the sounding cable. The lost pick axe slowed them significantly in establishing routes.

Leffingwell observed that the barometer had been holding steady since they left land for the pack ice. But by Sunday, April 14, the temperatures were warming up to -13° at 9:00 a.m. and -11° at 2:00 p.m. With warmer temperatures, there was more water, softer snow, and their feet were constantly wet. It was early in the year for the pack ice to break up. Again and again they approached cracks in the ice and wide lanes of open water. Two lanes of open water stopped them that afternoon. One ran north-south while the other ran east-west. At least they could see the older ice beyond. They often camped when they had put in a good days work or arrived at an obstacle. It provided them with an opportunity to do some reconnaissance for a route.

Leffingwell calculated a 10-mile difference in calculations using dead reckoning and the latitude observations. The party decided to wait for another day to obtain further observations and calculations on their drift. The following day yielded surprising and disappointing information. The lane they had camped near had opened overnight to about 300 yards. They did not detect any movement of the floe but the wind picked up. The winds prevented them from launching a raft, one they had fashioned from a piece of canvas wrapped around an 11-foot sledge, and paddled using a shovel. Mikkelsen had taken two observations for longitude and the results yielded the same information. They were rapidly drifting to the west and northwest! In the course of three days they had drifted about 20 miles in a north-northwest direction, further from Alaska, and headed for Siberia. Reality set in and Leffingwell was clearly concerned when he wrote: "Very bad, must cut S [south] tomorrow as M [Mikkelsen] & S [Storkerson] fd [found] lane runs that way instead of SW. 2-3 fresh bear tracks last night, one few hundred yds of camp. Lane widening still."[26]

They arrived at new ice and more water, and it was snowing. Leffingwell described their travel on the thin ice as unsafe and risky. As they crossed over one section, the ice bent more than they had observed so far. If it had broken, the consequences would have been deadly.

On April 19 they arrived at a pressure ridge that varied from eight to fifteen feet in height and consisted primarily of eight foot high blocks. While they were cutting a road for the sledges, the ice started moving one foot for every 20 seconds. Leffingwell wrote that the ice moved 50 feet before stopping. They finished digging the road and crossed over to settle into a new camp at 3:40 p.m. Another pressure ridge loomed ahead of them. The trio had a layover day and calculated the drift. Winds were coming out of the west and southwest but they were relieved that they were not drifting east. They celebrated and Leffingwell wrote: "Pleasant day making plans for next year, M [Mikkelsen] to Banks Ld [Land] from Dem Pt. [Demarcation Point]...Pem[mican] 6pm. Tea & jam 8pm. 5 meals instead of one as of old and primus going all day. Passed one of the most pleasant days in camp that I remember for long time singing, planning etc."[27]

On April 21, Leffingwell approximated that they had passed over 50 cracks, lanes, and ridges after digging out their sledges that morning. Their estimated latitude was 71°16' north with a longitude of 148°07' west, or by dead reckoning a longitude of 148°15' west. The next morning Leffingwell stayed in camp, reorganized sledge loads, and took photographs while their sleeping bags, socks, and other items dried in the sun. Mikkelsen and Storkerson took a sounding and were surprised when the result was 62 meters (372 feet). What they learned was that 2.7 miles north of their camp was the edge of the continental shelf. They headed north to establish the profile of the bottom they had not previously established.

When they reached a wide open lead the following morning, they had to stop and rig up the raft for the first time. Mikkelsen was the first across with a load and rope. Using the rope between them, Leffingwell and Storkerson could then relay the raft back for another load. They made six trips and were pleased with how well it worked. Acclimated to cooler temperatures, when the temperatures rose to 32°F, they stripped to undershirts while they worked. At times they opened the tent door because inside was uncomfortably warm. Due to the warmer temperatures Leffingwell modified his diet to reduce fat intake by eliminating a pound of butter per day, and increased the pound of milk to two pounds.

By Monday, April 29, Leffingwell calculated that they were about 25 miles from the lagoon where they had started. The route ahead looked promising and easier than what they had just come through.

Hopes for easier traveling did not last long.

They struggled with hauling the sledges, cracks in the ice, open lanes of water, and soft snow, often sinking to their knees and occasionally their hips. They were now down to three dogs pulling each sledge. Leffingwell wrote: "At 3:30 reached high hummoc...Lat at noon 71°00.6, one min. farther N. than DR [dead reckoning] for first time over estimated but in this going not to be wondered at. Hoped to be thro this bad belt by tonight but looks worse ahead. Vision of 3 days to land all gone. At rate will be 8 days, but it cannot last...Feeling discouraged for first time since start. Luckily have plenty of food (months) and plenty of seal or else it might be serious."[28] Leffingwell may not have realized that, given the early breakup of the pack ice and the remaining distance to the north coast, they were in grave danger.

They developed an alternate backup plan that called for abandoning sledges and traveling on foot. Interestingly, attitudes quickly changed once they settled into a comfortable camp. "All troubles over when in camp (5pm) and eating 2 lbs ration of milk & pipe after only a few smokes left. Must reach ship soon!!" wrote Leffingwell.[29]

Conditions were once again looking grim by May 2. Leffingwell and Mikkelsen scouted potential routes from a higher vantage point and saw nothing but water and a broken ice pack. Travel would be excruciatingly slow unless temperatures dropped and freezing occurred. The fog returned and visibility faded. They took a sounding and were encouraged that the water was shallower than the preceding day. But they were still rapidly drifting to the west and northwest, further from the coastline.

They started the following morning about 6:30 a.m. Leffingwell observed:

> Can't use boat in this stuff & can't cross with sleds. Must either abandon outfit or plug along as we do. Sding [sounding] at noon 32 m &

> S drift gratifying, as water S of us being closed up...Certainly we are glad that we are not farther from land. What if we had not turned back from 72°!!! Would be W of Pt. Barrow & make Wrangel Island after all. Sd 32 m...M [Mikkelsen] is walking up & down waiting for sun to get another obs...Ice closing in, in lane & drift S SW now, feel that if we can reach beyond water sky to S we shall be on fast land ice.[30]

Later that afternoon, the block they were on started moving before they were able to cross it. Leffingwell calculated its movement as one foot west every five seconds. By that afternoon they camped on a paleo floe surrounded by hummocks that reached to 18 feet in height.

Leffingwell described the open water as a "regular ocean" 250 yards south of camp. Given the circumstances, they had to wait and see where the floe took them, hopefully up against another floe so that they could continue south. The party had not been able to make observations for days so they were unable to pinpoint their location, but they took soundings that indicated a depth of 30 meters and a rapid drift to the west and northwest.

Mikkelsen calculated they were traveling about one mile per hour on the ice floes. At that rate Leffingwell figured it would take five to six days to reach Point Barrow and two days to Pitt Point. Leffingwell's dry sense of humor prevailed: "Lucky we are near shore...fine way of travelling. Wish we had map of Herald & Wrangel Ids etc. May come in handy."[31] (Leffingwell was making reference to the ill-fated North Pole attempt by the George Washington De Long Expedition of 1879. Their ship, the *Jeanette,* passed by Herald Island, was caught in the pack ice in the East Siberian Sea, and eventually sank.)

On May 6 they were able to leave their ice floe; crossed rubble and rolling ice to a larger floe. It was slushy, they were wet to their knees, and Storkerson's sledge overturned. Once they righted it, it slipped and knocked Leffingwell into the water up to his neck. Leffingwell was quickly warmed and later wrote: "Just as we got 3 sleds on last cake, 50 ft, it started to separate from other side. M [Mikkelsen] got across & threw us line & we got sleds onto 2 small cakes 12 ft."[32] They managed to reach a larger floe, but they were very lucky. The fifty-foot floe that they had just left cracked in a couple of places and was now 150 feet away.

They estimated their latitude as 71°13' north. At 2:00 p.m., Mikkelsen and Storkerson calculated that the ice cakes in the south lane were moving east at a rapid rate of one meter in four seconds! But the group was moving west and when Storkerson sounded the depth at 6:00 p.m., he found 29 meters, and they were drifting rapidly west and northwest. They each took a four-hour watch for opportunities to cross south, and in anticipation of traveling much further west, Leffingwell predicted that they would soon see whalers at Point Barrow.

Storkerson called them at 11:50 p.m. to let them know that they had come up against another floe moving rapidly. They broke camp and left within ten minutes. It was tricky because the ice was breaking into small blocks. They crossed and then reached a stretch of "very bad rubble" where Leffingwell's sledge overturned and spilled gear. They camped at 12:30 a.m. on a floe that had attached itself to shore ice. The sounding showed 24 meters, and Leffingwell knew they were safe. He and Mikkelsen stayed awake talking until 3:00 a.m., jubilant over having reached shore ice.

Storkerson's sledge was abandoned, which enabled Mikkelsen to determine the route ahead for Leffingwell and Storkerson with two remaining sledges. The weather was overcast and prevented them from determining their geographic location. They had not seen the sun for over a week.

The weather improved and travel became easier. Leffingwell wrote: "Knowledge that we are out of troubles and not drifting very comforting. Left eye smarts still."[33] Leffingwell had again suffered snow blindness for several days and was in great pain. As they traveled, he fell often, though Mikkelsen wrote that he never complained.

On Sunday, May 12, they were surprised to find an Inupiaq family camped on a sand reef at what turned out to be Jones Islands, located about 86 miles west of Flaxman, and directly south of their westernmost location while on the pack ice. They spent the day visiting, exchanging gifts, and drying their foot wear.

They camped on a sand spit west of Cross Island after a full day of travel the next evening. They were elated with their progress and Leffingwell wrote: "Due to the warming up we got in native tent. New life from contact once more with Earth. Also from great amt. of nicotine absorbed. Looking forward to dried fruit & sugar tomorrow at [Flaxman] Island. Too much fat & too little sugar in ration for last month."[34] Leffingwell was overly optimistic in reaching Flaxman Island the following day. They camped near Pole Island because of snow, poor visibility, and thirteen hours of travel. The snow was soft, the walking tough, and it was very slow going.

On Wednesday, May 15, after nearly two months, they saw the masts of the *Duchess of Bedford* through binoculars. And they were sighted by Fiedler and Howe, who traveled out to greet them. They arrived at Flaxman Island at 6:30 p.m. and had a large meal of caribou. Their journey on the pack ice was finally over.

Upon arrival at Flaxman Island, the team learned that they no longer had a ship. The vessel had been severely leaking for many weeks and was sinking. With the help of Inupiaq families, Howe and the crew had managed to get the remaining provisions to the island. They dismantled the ship for its timber and built a cabin on the island.

News Reports

Vilhjalmur Stefansson, the expedition ethnologist who had traveled via the Mackenzie River to its delta and overwintered there, arrived at Flaxman Island while the three men were out on the pack ice. Leffingwell, Mikkelsen, and Storkerson had not been seen since they left in mid-March, and word had been sent out via the RCMP at Herschel Island that the three had been lost and that their ship had sunk. Mail transport was slow and access to telegraph limited, so it was September 1907 before the news reached their families. Newspaper headlines blared "Arctic Explorers Frozen to Death," "Pole Hunters Vanish and Their Ship Sinks," and "Lost in Artic [*sic*] Ice: Three of the Anglo American Expedition Gone." Roald Amundsen was quoted as "'hopeful" the party had survived the sinking of the *Duchess of Bedford*. Leffingwell's mother remained optimistic about their well-being and told a reporter that just because she had not received a recent letter, she did not believe they had perished.

When Leffingwell and Mikkelsen learned months later of the false news reports, they were dismayed at the anguish their families and friends must have felt. Mikkelsen was particularly concerned about his mother, who had recently endured other losses. He traveled to Herschel Island and quickly sent word the party was alive. Mikkelsen also signed a contract with Storkerson for another trip on the pack ice to continue the search for land. Mikkelsen had asked Leffingwell to join him for another pack ice trip to follow the continental shelf edge both east and west, and continue the search for land. Mikkelsen later wrote of Leffingwell's response:

> "No, Miki," he replied gravely. "I knew you would ask that and I have thought and thought it over many a time. But I can't. Your work is on the ice, mine on land, and I can't give mine up. Also I no longer believe in the land out there, and I hate the pack-ice." That was no news to me, and I could sympathize with his point of view: his work as a geologist, his very future perhaps, depended on what he could find in the mountains there to the south—not on the hypothetical land out there in the north. And when he no longer believed in that land's existence, there was no expecting him to sacrifice more for it than he had already done. And besides, he had long since said that he wouldn't go—even before Storkerson had said that he would.[35]

Leffingwell's curiosity turned south to the mountains. Leffingwell ended his journal by writing: "I am off in couple of days with Ned [Arey] into Mts, so have to hustle. Can't sleep."[36]

The trip on the pack ice had pushed and pulled Leffingwell to his limits. He looked forward to working in an environment that was more suited to his training and interests. But his decision would not sit well with the American Geographical Society and the Royal Geographical Society. After all, determining if there was land north of Alaska and locating the edge of the continental shelf was the basis of their support, not a geology reconnaissance to the south. The societies expressed their displeasure to Leffingwell, but as his father had funded over $8,000 for the expedition, the issue was dropped.

Even if the explorers questioned their own work, the expedition made an important contribution to the understanding of the Arctic Ocean in northern Alaska.

CHAPTER 5

The Okpilak: "Place of No willows"

Turning his attention from the pack ice south to terra firma, the mountains beckoned Leffingwell. His work would now involve day after day of travel in a variety of conditions, careful observation and recording of the geology, analyzing and refining his knowledge, then pulling the information together by constructing maps. Leffingwell would add photographs and sketches to his field notebooks as further documentation.

Okpilak Lake, looking south toward Mount Michelson, 1991. *Janet Collins Photographic Collection*

While Leffingwell and Mikkelsen were on the pack ice, the *Duchess of Bedford* had been abandoned as it fell apart at the seams. Leffingwell's gear was scattered and time was needed to organize it. By the evening of May 17, 1907, Leffingwell, Arey, and Gallagher started their dog sledge journey from Flaxman Island to the Okpilak River drainage leading to the mountains. Evening was a fine time to depart, for in May the Arctic coast of Alaska is light for 24 hours a day. They started about 6:00 p.m. in thick fog and camped about 2:30 a.m. on a sand barrier island off the coast. Snow still covered the ground but conditions were changing rapidly.

Although Leffingwell and others developed different styles of tents, the Inupiaq dome tent was clearly the warmest and most comfortable. About twenty willow branches, eight to ten feet in length, were bent and then crossed over, making the base strong enough to withstand high winds. Sometimes the willows needed to be lashed if they had not been bent properly. Two layers of cloth over the top of the willows provided further insulation. A sheet iron camp stove was set in a corner and the stove pipe was vented through an opening in the canvas ceiling. In winter, they added fur skins to the roof and floor, and tamped snow around the perimeter to hold the canvas cloth in place.

Summer travel was possible at all hours, but changing one's personal clock to "night" was not always easy. When the weather was clear and the sun out, the tent became hot very quickly. Such was the case as they slept into the next morning, and Leffingwell stripped off blankets and his attege (fur lined parka). They prepared a breakfast of fried caribou and broke camp by 6:30 p.m. They stopped to camp about 1:30 a.m. east of Collinson Point. The weather was overcast and foggy, and Leffingwell described it as "not cold." They gathered enough wood to start the stove inside the tent, and Leffingwell noted that the heat removed the stiffness from his bones. He wrote: "Beginning to enjoy life again. Turned in 6 am."[1]

When they awakened at 3:00 p.m., winds of 15 to 20 miles per hour had risen and snow was drifting. The temperature hovered around -4°C (about 25°F). They continued on and Leffingwell noted that both the Sadlerochit River and Hulahula River were in flood. By the time they reached Arey's cabin on the Okpilak delta, Leffingwell wrote that the Okpilak was not yet flooding. The weather cleared and the mountains to the south were beautiful. During the day, pools of water appeared on the tundra as the weather warmed.

Leffingwell was highly organized and focused on the details of his trip. He made lists of scientific gear, camping and cooking gear, food for the party and food for the dogs, and clothing. He recorded the weight of each item. Over the course of his many journeys, Leffingwell recorded daily weather conditions, barometer and temperature readings, time and distance traveled, sightings of birds and wildlife, what they ate, and descriptions of their camps. Along the way to the Okpilak River delta, Leffingwell lost an important piece of his rifle, the front sight, which he repaired by cutting a nickel in half and mounting it on the rifle. His problem-solving skills would be tested time and time again during his Arctic years.

Along the coast Leffingwell observed and noted the diversity and number of birds: pintail ducks, owls, plovers, gulls, larkspur, snow bunting, and skuas. He wrote and sketched while Gallagher checked traps and hunted. Gallagher returned with five ground squirrels and a

Packing up the Okpilak River, 1907. *USGS Photographic Library, Leffingwell Collection*

ptarmigan, mainstays in their diet, supplemented with granola, bread, or boiled rice.

The distance covered each day varied widely depending upon traveling conditions, availability of wood, interesting geology, and preferences for campsites. After camping on the west bank of the Okpilak River, they left the delta and started their journey south. They traveled at night because the snow was firmer. Leffingwell wrote of soft snow, bare ground, and tough travel.

After sleeping about twelve hours they woke at 8:00 p.m. on Friday, May 24. They ate a cold breakfast because the Primus stove was plugged. They departed at 11:15 p.m.; their sled heavily loaded. As they walked, they frequently broke through the crusted snow. They continued slowly, stopped frequently, and established camp 3 at 5:00 a.m. The trio had passed the first of the foothills on the east side of the river and Leffingwell observed the geologic strata and wrote of seeing a Lapland longspur, a common songbird of the Arctic tundra.

Camp 3 was located on the west bank of the Okpilak, and there Leffingwell wrote that he was stiff and sore with fever and chills. His illness stopped travel for several days, then Arey and Gallagher continued ahead in the rain looking for firewood and established a new camp up valley. The next day Leffingwell had sore muscles and bones but his appetite had improved, so he set off with a throbbing headache and hiked a couple of hours to the next camp. Arey and Gallagher had set up the tent so Leffingwell crawled in. After sleeping a few hours he awoke feeling fine, washed up, cut his hair and beard, and ate some lunch.

As they traveled south, Leffingwell made a geology reconnaissance. At times warm winds blew from the south, but more often blew from the northeast, the prevailing wind direction.

On Tuesday, May 28, he wrote:

> Off 12:00 this am. Pools crushed over snow held in most places. Felt pretty good & helped load up etc. Had to take to tundra, snow tired me. Good going when struck gl [glaciated] area. Camped 3:30. Warm balmy breeze down valley. Lovely weather....Feeling elated to be on gl [glaciated] area again. ... Saw Lynx. Saw small bird fly up in air all time uttering single cry, then down ≤ [like] night hawk. Small purple flowers near tent. Gal saw butterfly. Clouds over mts [mountains] but clear overhead & to N. Can see crevasses in gls [glaciers].[2]

Leffingwell's geologic observations and descriptions were detailed in separate notebooks. Basic and necessary details and chores appeared to consume a good portion of the party's time and energy. The following day, as Arey and Gallagher continued moving loads upriver, Leffingwell gathered firewood, found a fossil in the river bed, stretched his uncomfortable hob nail boots by wearing them, and sewed a pack sack for Poyuk, one of the dogs.

Okpilak River valley, 1991. *Janet Collins Photographic Collection*

They were close to the mountains and Leffingwell began detailing the geology as he moved slowly upriver, his primary interest being Pleistocene geology and glacial processes. Pleistocene is a more recent epoch, about 11,477 years ago to 1.8 million years ago, within the Quaternary Period of the Cenozoic Era, the most recent era of the four eras defined in geology.

They had clear beautiful weather. Temperatures reached 16°C to 20°C during the days. His legs were strengthening but the straps and weight of his canvas pack were uncomfortable. Leffingwell noted one recessional moraine which he referred to as the last one outside of the mountains and also wrote of the stratigraphy.

On May 31 he climbed a hillside and discovered an outcrop of sandstone, quartzite, and shale, but no fossils. By the time he returned to camp he was tired and his knees felt weak. Gallagher and Arey shot two caribou and returned with a large quantity of meat.

He took compass bearings (readings) on significant physical features, then plotted them on paper on the plane table. He was then able to establish a series of triangles based on defined locations, or triangulation stations. The defined locations provided the basis for map accuracy. Clouds moved in and, unable to take angles, Leffingwell returned to camp, leaving his equipment temporarily behind.

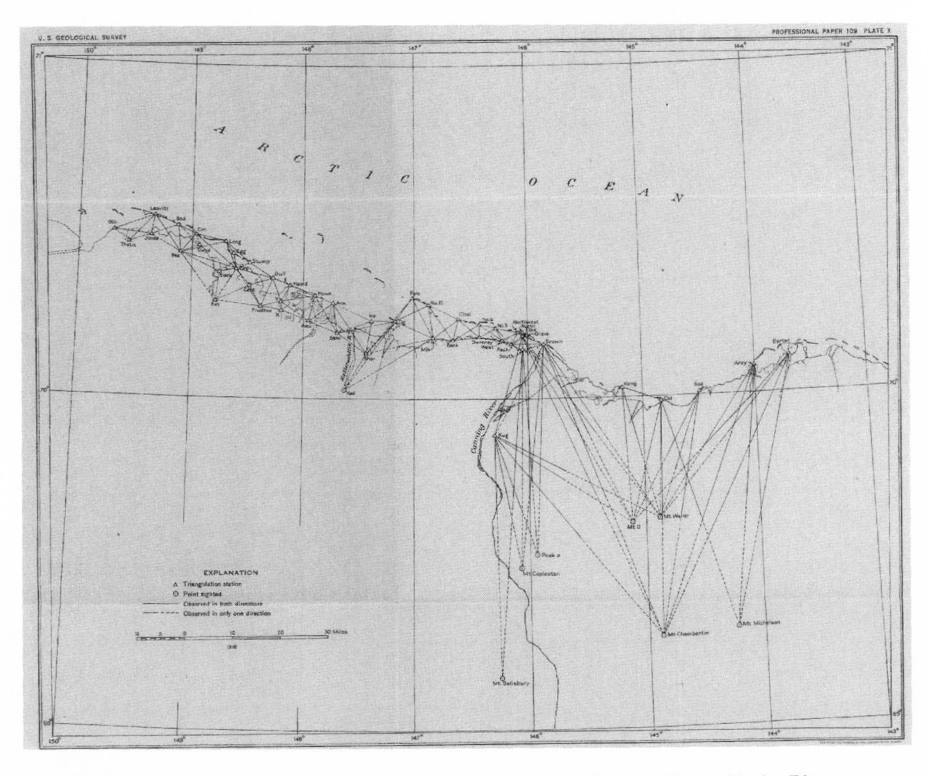

Triangulation stations along the Arctic coast adjacent to the Canning River, Alaska. Plate 10. *USGS Professional Paper 109*

Leffingwell's feet were so sore from the hob nail boots he stayed in camp the following day and packed to travel upriver. He was up at 9:00 a.m. the next morning and back to work. His legs were feeling stronger, so he returned to the site he had visited two days earlier, retrieved his equipment, and started taking angles. Leffingwell also constructed a "signal," although he did not identify whether it was a rock cairn or another type of material. Signals at various locations were essential for him to establish the triangulation network. He would take sights, or angles, on the signals from other locations miles away to the signal he had recently established. It was 8:00 p.m. before he started his return; he arrived at camp two hours later in heavy fog.

Leffingwell traveled upriver looking at the geology, and when he reached camp 7 near some small hot springs, he noted the granite bedrock in a nearby alcove. Granite's resistance to weathering helps explains the high terrain of that section of the Brooks Range. He spent the evening cracking rocks and studying their composition. Snow fell the next morning, June 8, and the tent leaked. By noon the fog was gone and Arey and Gallagher hunted up river. Their meat supply was gone. Leffingwell started out to take angles, but the fog returned. He returned to camp and wrote up his geologic notes. Temperatures were cool, 3° at 2:00 p.m. Leffingwell crawled into the tent; he was cold. Later that evening he hiked above camp and collected granite specimens.

The following morning he tried to take angles but the fog and clouds made his work impossible. They continued up the Okpilak Valley toward the forks near the river's headwaters. The clouds and fog lifted briefly late in the afternoon and Leffingwell noted many high mountains and glaciers up both forks.

Wednesday morning, June 12, was partly cloudy, 12°C at 9:00 a.m. and no wind. Weather in the Arctic is rarely consistent, so it was no surprise that later the same day, there was sun, then rain. It rained hard during the night and the river rose. Leffingwell forded the river to take angles, once again working to map the surrounding peaks. He described the West Fork as white mud, a glacial flour, or ground glacial rock. Footing was difficult as he crossed, and he was wet to his knees. Floating pieces of ice and freezing water must have felt like needles to his skin.

Hours later the river was even higher and he could barely stand against the current. As he recrossed the river, Arey had to throw him a staff. Thankfully, a change in the wind direction and cooler temperatures meant there were fewer mosquitoes. Gallagher and Arey had killed a ram, so they had mutton for the evening meal.

Over the next several days, Leffingwell continued hauling his plane table to establish new triangulation stations for mapping the surrounding terrain. He collected rock samples of gneiss, a type of metamorphic rock, and granite specimens, made notes about the glaciers, took photographs, and wrote up his notes in the evenings. His feet suffered in the hob nail boots. He had been accustomed to wearing kamiks, soft and comfortable Inupiaq moccasins. Daily temperatures were increasing, now 16°C to 18°C. On June 17, he again crossed the river but "got swept off feet in crossing & wet all except head. Luckily arms caught on bottom & I got out quickly...Lost fur coat had it over shoulder, just took it off. Stripped all but under shirt & sat in Ned's bag for supper then single blanket all night."[3]

When rivers were significantly swollen by rain, there was real danger of being swept away when crossing. Leffingwell was much more careful after his close call.

The snow level dropped to 1000 feet and his clothes were still damp the next morning. They headed further up valley and camped at the forks that he had earlier observed. The weather was warmer and mosquitoes thicker. The following day it rained and Leffingwell spent the day reading in the tent. He understood that when weather was poor, it was best to wait it out.

Of the four river valleys Leffingwell would explore until 1914, the Okpilak River drainage he described in U.S. Geological Survey Professional Paper 109 was the one "most marked by glaciation." His descriptions were highly detailed. He described the upper 20 to 30 miles of the Okpilak as a deep glacial trough, a U-shaped valley created by a glacier flowing down valley. Near the headwaters, Leffingwell had primarily explored the West Fork of the Okpilak. He wrote that its source was a glacier that extended six to eight miles. Leffingwell then described the East Fork as narrower and shorter. As he traveled up and back down the Okpilak River valley, he determined that 2,000 feet of ice had covered the valley floor at the bend in the west fork of the headwaters. He described glacial processes; areas of recessional moraines—moraines composed of material deposited as the glaciers retreated. Leffingwell indicated the terminal moraine was not exposed, although Alaska geologist Gil Mull has indicated that it is significantly north of the mountains.[4] Leffingwell also described the talus slopes, the lateral or side streams, and rock types that included granite and schist.

Over the next two days the weather was varied—rainy, cloudy, foggy, sunny, and clear. Arey and Gallagher hunted and Leffingwell collected rock specimens if conditions allowed. When it rained, Leffingwell worked on the map until the leaky tent forced him to stop.

By Sunday, June 23, Leffingwell had completed mapping along the upper Okpilak, so they headed down valley toward the coast and previous campsites. The temperatures increased and so did the mosquitoes, so Arey and Gallagher made head nets. They continued down valley and Leffingwell cached some rock specimens because of weight. The sole of one of his boots separated, so he tied them on with cord, but they kept slipping off his feet.

On Monday night they established camp 15—which was camp 7 on the trip up valley—and stayed for nine nights. The camp was near the hot springs and on the opposite side of the river near a place they referred to as Sheep Creek.[5] From the forks of the river near the headwaters, Leffingwell had noted the occurrence of granite down valley "to the north edge of the mountains." And he observed that near the northern edge, shattered and folded limestone overlay the granite.[6]

On June 26 the temperature was 20°C; the mosquitoes were becoming thicker and more annoying. Leffingwell's dog, named Verner, was not a local North Slope dog, and suffered. His face was swollen but the local dogs did not appear to be affected. Two days later, Leffingwell wrote: "Couldn't get asleep. Mosqs very bad outside & few in tent. Kept dogs on move all night & quarrelling & kept us awake. 5:00 am Verner stole sinew drying on stove pipe. Chased him around & couldn't sleep till 6:30am. Up 11:30 & big feed of mutton. Heavy clouds & rain above. Lower fog over us. Breeze & cold kept mosq down till 8 pm. Stew for supper. Did not work except to crack out fossils. Fire great comfort. Talked with Ned...Ate dried caribou all day."[7]

Leffingwell explored the mountains, hunted, hauled rock specimens and cracked out fossils. Some of rocks contained brachiopods, a small shell fossil. He also did some field sketching, took photographs, and when the weather was especially bad, did very little. On July 3 they headed downstream and spent five nights at the next camp. They gathered enough wood for supper and Leffingwell spotted a lynx about 100 yards from their camp. Lengthy stays in the various camps provided Leffingwell time to "take surveying angles" for his mapping project, and study the geology.

Leffingwell rarely complained about weather and conditions, but after having been out since May 18, he wrote on July 7: "Getting sick, tired of eternal rain, grub very monotonous & unsustaining unless one eats all one can cram down. Miss bacon, bread & beans, taking grub for 4 days to coast. Cached sheepskin, primus, oil can (1 filling left), specimens, a little corn meal, 10 lbs sugar, 2 large sheets drawing paper. Used stove 1 day here."[8]

By the next day his mood had improved. They continued toward the coast and camped 15 feet from the river in low brush and a grassy patch. He noted that it was the "best camp yet." The river was at least a foot and a half higher than their earlier camp at approximately the same location. They retrieved food from a previous cache and Leffingwell devoured freshly made biscuits for the first time in six weeks.

Early the following morning he "took angles etc & photos of mts." Within a few hours they departed camp, traveled 10 miles in six and one-half hours, and camped about three miles below what had been their camp 2. The morning had been clear over the mountains but foggy at the coast, a common occurrence in summer. Leffingwell's feet were sore, his legs chafed, his hips bruised by his pack, and his head and shoulders sore from the pack straps.

They traveled 13 miles in seven hours the following day. As they approached the coast, the wind blew about 20 miles per hour and brought relief from the annoying mosquitoes. Leffingwell had another miserable day, even without the mosquitoes: "Feet blistered by 2pm. & instrument box digging chunks out of spine."[9]

The following day the trio made it to a camp near the mouth of the river, after traveling nine miles in seven and one-half hours. It was a much easier day of travel. On a fork of the Okpilak on the river delta, they reached the boat of Terigloo, a hunter hired by Leffingwell. They hopped into the oomiak—an open boat made of seal skin stretched over a wood frame—and rowed the remaining distance to Arey's camp on what is now known as Arey Island. Leffingwell dressed in warmer clothes, made note of the cold (5°C and a 30-mile-per-hour wind), and observed that the pack ice was out about five miles. They pitched camp on a bank above the mud flats. Over the next few days, he read, hunted squirrels, and ate what he referred to as a "steady diet of beans." Arey's wife, children, and others arrived at the camp on July 16, and Leffingwell learned that Mikkelsen was not coming. He did not know that George Howe was on his way instead.

On the island, they needed fresh water. They dug a hole near the tent but the water was too salty to use for anything other than tea. Ice could have supplied it, but there was none. Some of the Inupiat went to the mainland for fresh water and to hunt.

One of the women made another attege for Leffingwell to replace the one he lost when he fell crossing the Okpilak River. He continued his work taking and plotting angles and completed mapping the island.

Howe and Joe Carroll arrived with a seal boat (oomiak) to pick up Leffingwell and return to Flaxman Island on Monday, July 29. Howe also brought Leffingwell mail from the previous year. A storm blew in and strong winds prevented their departure. Finally, on August 2, they started for Flaxman. Leffingwell wrote:

> Fog came up just as we started & got stuck on mud flats in breakers which splashed over us. Poled off ok & set sail. Wind on quarter. Boat road long swells nicely. Fog lifted about 11am & saw shore to whaler. Headed out but fog so squared away & did not sight land till fog lifted & could see pole on Coll [Collinson] Pt. Wind freshened & white caps so put around & camped half mile above pole. Only few pieces of ice seen. Cleared off fog good 4pm as we reached Pt., Joe cooked supper, eider duck stew, over campfire & we sat around till 9:30. Very pleasant.[10]

They had a layover day near Collinson Point due to the rain and thick fog. Leffingwell spent the day surveying. The following day crew member Thuesen and an Inupiaq, Alikok, arrived from the east in a boat with a message from Mikkelsen. Storkerson was dangerously ill, so Howe left with them to care for him. Storkerson decided to leave for home and broke his contract with Mikkelsen. Without Storkerson or Leffingwell, Mikkelsen would be unable to return out on the pack ice. The Inupiat were wary of the pack ice and finding help appeared impossible.

The party then consisted of Leffingwell, Carroll, and two Inupiat in the small boat with gear and three dogs. They continued west and camped east of the Kugruak (Canning River) mud flats. From there, they traveled from Brownlow Point, reached the eastern end of Flaxman Island about 3:00 p.m. on August 5, and arrived at the cabin later the same evening. Leffingwell's first scientific journey into the mountains had ended.

While Leffingwell explored the Okpilak River drainage, Mikkelsen had traveled to Herschel Island on a whaling ship. During that trip in early August he learned their party had been reported "lost" on the pack ice. Mikkelsen wrote letters seeking financial support for another trip out on the ice. He hoped to find someone to accompany him and to acquire some additional sledge dogs. Given the reports of their demise and potential for danger, no one appeared interested.

Home Again on Flaxman Island

Awaiting Leffingwell at Flaxman Island was a box from his family in Illinois containing magazines, candy, cigars, and cigarettes, but no mail. Over the next few days, Leffingwell arranged gear, and repaired and adjusted the chronometer and altazimuth. He lamented wasting the good weather. Within a two week time period, temperatures varied from 2°C to 10°C; the weather was clear, the winds were mild and out of the northwest, and the mountains were visible. He continued drawing plans for a canoe and made a frame for bending sticks for a tent. He wrote letters for an entire day. Howe helped him erect a pole for astronomical observations on the sand spit north of the island. In the evenings, he accompanied Howe to Brownlow Point, where Howe provided a medical clinic for the local families, treating ailments from stomach aches to tuberculosis.

The crew members were also busy with projects. They collected moss for the roof of the cabin. Thuesen built shelves in the cabin and worked on the tent sticks, Fiedler worked on the tent cover, and Storkerson worked on weaving dog harnesses. Hickey cooked. As the month passed, Thuesen and Fiedler sawed wood for the boat. From time to time they had visitors for dinner and played the phonograph for entertainment.

August 19 the crew attempted to steam and bend the eucalyptus wood to form the ribs of a boat, but the ribs kept breaking; same with the stern posts from the ship. Leffingwell suggested Thuesen steam some oak planks the next day and found that they bent "beautifully."

On August 22 Storkerson again advised Leffingwell that he was going home when the rest of the crew did. Storkerson would not travel with Mikkelsen on another pack ice trip in search for land to the north.[11]

During this time Leffingwell maintained the household, including baking bread. While filing and sharpening a butcher knife, it fell point down on his foot, cutting to the bone but missing his ligaments. When Howe returned later in the evening, he dressed the wound.

On August 25 Captain Hoffman and the whaling ship *Jeanette* arrived from the east. It was common practice for whaling ships to announce their arrival with gunshot or the ship's whistle as they neared settlements. Mikkelsen, who was at Herschel Island, told Hoffman to help himself to their storage of meat, so Hoffman came ashore and stayed for dinner. Leffingwell supplemented the meat with a box of lemons. Hickey returned south to the lower 48 with the *Jeanette*, while Thuesen, Fiedler, and Storkerson waited for the arrival of the whaling ship *Narwhal*.

Around breakfast time on August 28, the *Narwhal* announced its arrival with the sound of a whistle. The seas were rough and it took hours before the ship was able to lower a dinghy and it was 5 p.m. before the boat reached the west end of the spit. "Mail for Dr. H & me from Home & note from Leavitt [George Baker Leavitt Sr., captain of the *Narwhal*]... Mist up & when reached cabin after 9:00pm couldn't see ship. Cold supper & turned in. Left boats on w spit."[12]

The crew of the *Duchess of Bedford* departed aboard the *Narwhal*. Leffingwell remained to explore and map the mountains to the south. Shortly after the crew's departure, Leffingwell observed the fall migration and the gathering flocks of snow buntings, phalaropes, long-tailed ducks, brant, arctic terns, skuas, and glaucous gulls. Some departed, others waited a bit longer. Snow fell late in the afternoon on August 31 and the next day, Leffingwell noticed ice in the buckets for the first time. As he looked south toward the mountains, they were covered with snow to their base. It was now dark by 10:00 p.m.

Mikkelsen Returns From Herschel Island

On the evening of September 2, a ship's whistle announced its arrival. Leffingwell showed a light outside so the ship knew someone was at the house. The ship had dropped off Mikkelsen, who arrived at the camp about 2:00 a.m. They talked for hours and decided to retrieve Mikkelsen's gear from the point. On September 4 Leffingwell wrote that Mikkelsen had decided to leave for home.

The next several days were busy. Leffingwell overhauled the sledges and whittled new tent stakes, while Mikkelsen worked on a new darkroom where bunks had once served the crew. The sun set at 7:15 p.m. on September 7, and by then Leffingwell had completed writing about his work for the past year and "turned toward planning future work."

Mikkelsen and Shugvichiak headed for Collinson Point on September 11 to find Little Uyarak, an Inupiaq, to see if he would accompany Mikkelsen overland. While they were away, a strong gale blew at Flaxman Island and Leffingwell wrote that the "Bluffs undercut & many slumps where ground ice holds up is undercut over 10 ft in places."[13]

Leffingwell's first house, Flaxman Island, 1907. *USGS Photographic Library, Leffingwell Collection*

(Undercutting of the banks remains pronounced today along the riverbanks as well as around the islands.)

A few days passed, Mikkelsen was away, and Leffingwell offered a brief and rare glimpse into his inner thoughts. "Hard to work when alone, rather read & loaf. Spend hours every night going over programs of field work & field methods. Have to solve details myself as work is different from F.J. [Franz Josef] Land. Ordering some more draw instruments & paper. Cook used draw board for cover to water pail etc & other one spoiled in shifting from ship."[14]

It snowed and blew all day on September 16. Leffingwell developed photos in his new darkroom and baked bread. Later as he carried buckets of water from the pond to the house, he had to walk sideways because the snow was cutting his face. The next day, he noted the passing of his first year at Flaxman Island. Leffingwell layed moss over the roof for additional insulation the following day, and recorded that the moss made 3° difference in temperature inside the cabin. The roof moss was eventually covered with a wire net to secure it from the harsh winds. He caulked cracks to keep out the wind. Ice formed in the buckets on the floor.

Leffingwell thought of Mikkelsen and Shugvichiak as a gale with winds in excess of 25 miles per hour blew from the northeast. He busied himself by cutting and hauling ice to the cabin, using the clear ice to cover the windows and provide further insulation. Winter was once again approaching. One evening he went to bed by 6:30 p.m. just to keep warm.

The last two weeks of September Leffingwell traded sacks of flour and other food with Inupiaq families on Flaxman Island in return for animal hides. He was busy teaching several of the Inupiat the English alphabet. His interactions with nearby Inupiat were frequent, often daily. They exchanged gifts, shared food, traded, and helped each other as needed. Toklumena, Shugvichiak's wife, mentioned to Leffingwell that she had seen returning hunters on the mainland. Ice conditions made it impossible for her to pick them up by oomiak and return them to Flaxman Island. When she was able to retrieve the hunters, they gathered at Leffingwell's cabin while he cooked up a "big feed."

The weather warmed again and by September 27 snow had melted and the ice disappeared. When two Inupiaq families returned to Flaxman

Island on September 30 from the east, they reported that they had not seen Mikkelsen or Shugvichiak, who had left for Collinson Point on September 11. Two weeks later, they were long overdue. Leffingwell wrote: "No one has seen Sagi [Shugvichiak] & M. & I am afraid they are gone. Prob blew out to sea in SW gale day after they started. May have got into pack & may show up, perhaps at Ned's tho M ought to be back by now if he is alive. Toklumena is worried but has not given up hope yet."[15]

Fortunately they did not have to wait long. On October 2, they heard a gunshot from Brownlow Point. An oomiak with six men traveled the three to four miles from Flaxman Island to Brownlow Point and found Mikkelsen, Shugvichiak, and a Swede by the name of Eskal waiting. They had managed to reach land during the gale, and met up with Arey in the mountains, where they hunted caribou. Leffingwell wrote: "When Sagi [Shugvichiak] came home Toklumena had tea & flapjacks ready. He simply shook hands & sat down to eat. This after being given up from the dead."[16]

The Anglo-American Polar Expedition Comes to an End

Mikkelsen had already wavered about leaving for home or staying. He considered another trip onto the pack ice with Eskal, but by October 5 had once again decided to leave. Inupiaq families sewed clothes, including a new attege for Mikkelsen to wear on his journey. Mikkelsen packed food and prepared the sledge for departure. His route would take him west to Barrow, then along the coast to Nome, east to Fairbanks, south to Valdez, and then south by ship.

Mikkelsen had decided to leave, but Leffingwell would stay and continue mapping and surveying the coast line and the interior, including mountains to the south. While Mikkelsen prepared for his departure, Leffingwell began preparations for his next trip, up the Canning River. He sewed a tent, made two dog harnesses, added a new needle to his compass, sewed clothes, and prepared the sledge. Temperatures were falling and by October 15, it was -10°C. About noon, thirty Inupiat gathered at the cabin for a farewell party for Mikkelsen, featuring tea, an Inupiaq dance, and a meal of pemmican stew. Leffingwell danced a "two-step" to the phonograph and wrote that Mikkelsen presented gifts that included tobacco and calico fabric.

There is some question as to whether the Harvard anthropologist Vilhjalmur Stefansson was truly a part of the expedition. Invited to join what would be his first trip to the Arctic, Stefansson had traveled down the Mackenzie River and ended up at Herschel Island in the Yukon Territory, where he was to rendezvous with the expedition. When Stefansson learned that the *Duchess of Bedford* had become stuck in the ice near Flaxman Island, he decided to remain at Herschel Island and focus his work on the peoples there and around the Mackenzie River delta. Stefansson did travel to Flaxman Island while Leffingwell, Mikkelsen, and Storkerson were out on the pack ice, and again later that summer to retrieve some AAPE provisions in order to settle his accounts at Herschel. The AAPE had provided him with funds to travel north. Stefansson had little contact with Leffingwell and Mikkelsen, and his goals and work plan were performed independently of the expedition.

The Anglo-American Polar Expedition had ended. Leffingwell and Mikkelsen signed a new agreement on October 16, 1907. Leffingwell would assume the debt of payment to Howe and Thuesen. In return Mikkelsen gave his share of food and gear to Leffingwell. In addition, the agreement specified:

> 1. The previous contract bet[ween] us null & void.
>
> 2. M return to civilization. L remains to carry on the work of Expdn.
>
> 3. The entire property of Expdn. is turned over to L.
>
> 4. The $300 advanced on Thuesin [sic] wages and M's half of Dr. H. note for $600 are assumed by L.
>
> 5. Each is entitled to all the money coming from his own publications & lectures.
>
> 6. L agrees to permit M to use photos taken by L.
>
> 7. M agrees to take no steps which would prevent L's future use of above photos.[17]

The second point is misleading because the work of the expedition and funding by the American Geographical Society and the Royal Geographical Society in London had been to determine if there was land north of Alaska. Determination of the continental shelf so close to land had indicated that land north of Alaska was unlikely. Mikkelsen wanted to continue the expedition's mission but had chosen to leave because no one would accompany him on another pack ice trip. Based on the funding provided by AGS and RGS, there was no further work by Mikkelsen to be carried out.

The agreement was further complicated because half of the expedition had been funded by Leffingwell's father. In addition to determination of the continental shelf, Leffingwell felt a responsibility to his parents and himself to produce scientific work representative of his training and interests to justify the amount of money that had been invested in the expedition by his parents.

Mikkelsen continued preparations to leave and finished by 8:00 p.m. Leffingwell wrote, "signed new contract, wrote last word home, read till 1 am. Beautiful night, aurora, moon & stars."[18] Mikkelsen left for Barrow about 8:00 a.m. the following morning, with five dogs and a heavy sledge weighing about 650 pounds. The temperature was about -10°C. After Mikkelsen departed Flaxman on that beautiful quiet morning in mid-October, he and Leffingwell would not see each other again for forty-six years. Mikkelsen described the parting in *Conquering the Arctic Ice* and *Mirage in the Arctic: Explorations in Unknown Alaska*:

> Mr. Leffingwell said good-bye just below the bank. Words did not come readily to us, but we thought all the more. A moment we stood with hands clasped, then abruptly said "good-bye," a silent wish that everything would go well with the other, and the partnership, which we both had done our best to keep so smooth as possible, from which we both had derived so much pleasure and carried away so many pleasant recollections, was dissolved...Now and then I looked back at the black solitary figure, who in his turn was waving his hand at us, while high up on the bank we saw the familiar picture of our Arctic house and home. But then a point was passed, and house and ship passed for ever from view.[19]

As required, Mikkelsen submitted another report to the AGS before he departed the Flaxman base camp. AGS, in turn, wrote Leffingwell on November 8, 1907 expressing its concern about Leffingwell's "reputation with scientific bodies." It did not sit well with AGS or RGS that Leffingwell was unwilling to make another trip on the pack ice. The search for land and the continental shelf had been the justification for the expedition and funds provided by the two societies and their wealthy members. The societies were hoping for more discoveries, perhaps new lands, not the end of the expedition.

AAPE Accomplishments

In a report to the Royal Geographical Society Mikkelson recounted the expedition accomplishments. Of the total 533 total miles they traveled, 361 were on the pack ice. He included their drift calculations in his report: "We crossed the edge of the continental shelf which was found 71 12 N. Lat. and 148 24 W. Long. and travelled about thirty-five miles further north. During our ice sledging a strong westerly ice drift was observed in calm weather and with easterly winds but there was no set of the ice to the eastward even during heavy westerly gales."[20] Mikkelsen described travel conditions, ice and snow, and provided comments and recommendations about sledges, skis, snowshoes, and clothing based on their pack ice experience.

Mikkelsen and Leffingwell had already determined that the maps and charts of the north coast were inaccurate. Leffingwell planned to survey the coastline and Mikkelsen contributed to those efforts: "In the early summer of 1907, I went on a boat trip from Flaxman Island west to the Colville River and made a series of sketches of the coast line in that direction. Poles were erected at all the more prominent points to facilitate an accurate survey by Mr. Leffingwell."[21]

Their scientific work included meteorological and tidal observations. Meteorological observations included temperature, barometric pressure, wind speed, and weather conditions. Expedition members had taken tidal observations from the time they had arrived at Flaxman on September 17, 1906, until the end of the year. Mikkelsen noted, "The observations were taken hourly, and at high and low water during every ten minutes before and after the change."[22] Leffingwell made tidal observations at Icy Reef for three days and Howe had done the same for nine days at Pole Island. Mikkelsen indicated that all tidal and meteorological records were with Leffingwell on Flaxman Island. The tidal records were of particular importance because they addressed theories of ice movement and tidal actions.

Mikkelsen then wrote that Stefansson had traveled from Herschel Island to Flaxman Island and had accomplished some ethnological work at Flaxman Island during July and August of 1907.

Plagued by rough weather and conditions, Leffingwell and Mikkelsen had learned a significant amount about conditions and survival in the Arctic on Alaska's North Slope. They may not have accomplished what they had hoped—finding land north of Alaska—but they had determined the edge of the continental shelf, in itself a significant achievement. In spite of determining the edge of the continental shelf, both Mikkelsen and Leffingwell at one time or another continued to believe that there was land north of Alaska.

Dr. Chamberlin, geologist at the University of Chicago, wrote Leffingwell at Flaxman Island: "Your discovery of the edge of the continental shelf at a point so near to the coast is a contribution of high value. It will apparently lead to a reconsideration of the inferences based on the tides and similar phenomena."[23] Chamberlin had corresponded with Leffingwell about the Flaxman Formation, discussing theories and hypotheses, and was very supportive of Leffingwell's desire to explore and learn more about the geology of the mountains to the south. Chamberlin believed Leffingwell's work in the mountains would be a valuable contribution to geology in understanding their formation, structure, and glaciation. Leffingwell looked forward to that exploration, where his determination, flexibility, and resiliency would be tested time and time again.

CHAPTER 6

The Canning and Hulahula Rivers

Following Mikkelsen's departure on October 17, 1907, Leffingwell continued preparations for his exploration of the Canning River. He referred to it as the Koogoora, the Koogooruak, the Koogaruk, or the Kugruak, meaning "old river." Okalishuk helped transport loads toward the river to cache four miles beyond Brownlow Point. Leffingwell's work plan followed the template he had established up the Okpilak.

Leffingwell provided what medical assistance he could to nearby Inupiat, who shared the same symptoms: high temperatures, high pulse rates, and intestinal distress. His treatment consisted of a milk-only diet and a one-quarter gram dose of morphine. Alikok cut his large toe with an adze, so Leffingwell washed and dressed the wound with tobacco. Two days later, he returned to Brownlow Point to check on the sick. He acknowledged that travel between Flaxman Island and Brownlow Point consumed a good part of a day. Another family was sick, and he provided a concoction of chloroform and morphine as cough medicine, as well as food: milk, crackers, and two ducks.

Mikkelsen had hired an Inupiaq couple, Flavina [or Flaveenek], and Mamayuok, who with a young daughter arrived at the Flaxman base camp on October 19 from Herschel Island. Leffingwell made space for them and agreed to hire them for two years.[1] He was to provide them with a whale boat in exchange for their help. The family would help with scraping skins, sewing, cooking, and hunting. They were also good company. Leffingwell described playing the phonograph and building towers of poker chips with the little girl. He also worked on last minute items before his departure with Okalishuk. He sewed up a tear in the tent cover and experimented with different types of socks by wearing fur socks on one foot and two wool socks on the other. It was part of his character to constantly refine his preferences by experimentation to determine what worked best.

Leffingwell planned to leave on October 26, but the wind blew, the barometer dropped, the temperature was -6°, and snow fell. He spent the day reading and doing chores. The following morning he and Okalishuk departed with a loaded sledge, four dogs, and one month's supply of food. They made camp at 2:00 p.m. and cached their load after traveling eight miles. Winds blew from the west at 20 to 25 miles per hour, but they had the tent sticks up, the tent covers on, and a fire built within an hour. That evening gale winds blew in excess of 40 miles per hour.

They remained in camp the following day due to the wind. Okalishuk taught Leffingwell how to make a wind break against the snow and Leffingwell taught Okalishuk English. Winds increased the next day, and visibility was less than 200 yards. The outside temperature was -3°F but they were comfortable and warm inside. The winds subsided by late afternoon and they were able to see the mountains.

Leffingwell decided to return to the Flaxman camp after four days of being tent-bound by the weather. Provisions were low and there was little wood. Leffingwell and Okalishuk left the tent and their load behind and departed for the island on Wednesday, October 30. They reached the house about noon in a snow storm. Leffingwell returned to the camp the next day and dug out the tent and sledge. Leffingwell thought that perhaps a fox had managed to eat some canned dog pemmican so Okalishuk returned to the house for more pemmican and candles. They departed camp early on November 2 to head upriver. They traveled 21 miles in seven hours and camped at 2:00 p.m., after struggling with the sledge on glare ice.

They started early the next morning, and covered about 25 miles in seven hours. Leffingwell continued to struggle with comfortable footwear. He tried his "creepers" (similar to crampons), and found that his feet hurt when he wore them. He tried his kamiks but the ice cut into them.

Two challenging days later they reached Shublik Island, a conspicuous rock outcrop in the middle of the Canning River. Leffingwell wrote: "At 11:00 water everywhere, but got over onto thin ice. Held me but broke with sled but only 6 in deep. Got sled on top & ok for ½ mile when broke...again up to knees, sled & all. Lucky good ice only 20 ft away so cut dogs loose & boy tied all traces & threw me line while I sat on sled. Then I had to break ice to shore & we got her out. Froze fast from water dripping etc."[2]

Leffingwell and Okalishuk camped near an Inupiaq family and discovered that most of their gear was soaking wet, including clothing and scientific equipment. Fortunately, the camera had been spared. Leffingwell's kamiks froze to his feet and had to be thawed before he could take them off. That night the river rose, soaking the underside of their tent. They had placed willow branches beneath the skins in the tent, which protected them somewhat, but their sleep skins and atteges were wet. They spent most of the following day, November 9, drying clothes and wet instruments.

As the temperature warmed from -21°, it began to snow. Snow fell the following day and six inches accumulated. There were other Inupiaq families camped in the area for fishing. While Okalishuk fished, Leffingwell went for a walk and picked up a few rock specimens. One of the family members mended Leffingwell's boots and stockings, and in return, he gave them 30 pounds of cornmeal and some pemmican. He also cached some meat, butter, biscuit, and a net. Shublik Island became the turnaround point for the trip.

Leffingwell and Okalishuk started their return to the Flaxman base camp on November 14, but did not travel far. The river ice was thin and cracking. A gale coming from the east down the Ignek Valley blew snow in the air about 100 to 200 feet. They left the sledge and returned upriver to the group camp. On November 16 Leffingwell and Okalishuk again began their return to Flaxman Island. For three days they made good time; 21 miles in six hours, 24 miles in six hours, and 16 miles in four hours. But on the third day the wind returned and limited visibility. They struggled as the sledge slipped around on the ice. They decided to camp early and stayed an additional day. Temperatures ranged from -20°C to -25°C so it was no surprise that Leffingwell had to thaw his kamiks before he removed them. Even though the socks were frozen to the boots, his feet were warm. He wore two pairs of fur socks and stuffed grass in the kamiks and noted how warm the fur socks were compared to wool. On the final day of their trip Okalishuk shot a caribou and they set two traps for fox. They reached the Flaxman base camp in early afternoon on November 20. They had been away since October 31.

Back at Flaxman Island

Back in comfortable quarters, Leffingwell read for most of the first week and visited with nearby Inupiat. He traded food for animal skins. Shugvichiak brought him half of a bear that he and Flavina had shot. Leffingwell enjoyed the company of the Inupiaq family. He wrote: "Girl decorated my face & arms with endel [indelible] pencil & I gave her imitation tatoo on chin. Lots of fun. Also play 'bean porridge hot.'"[3]

In late November Leffingwell worked on the report and map of his Okpilak River trip. One conclusion from the Okpilak trip was that some of the mountains in that area reached about 9,000 feet, a fact later confirmed.[4]

On December 2 a gale blew between 35 and 40 miles per hour. The outside temperature was -32°, and the indoor stove burned at full capacity. Early that morning, before Leffingwell was awake, Flavina had discovered that the stove pipe was red hot. Leffingwell described the incident:

> I jumped up & grabbed breeches & dressed all but feet. Sent boy & man outside to dig away the sod and put snow on the pipe while I put a dish pan under and stood by with water in case the pipe should break or melt. Got fur clothes handy but left feet bare for I did not want to get gear wet and then get frozen if the house burned & we had to go outside. I did not realize that I could change out in the snow entry way. Got almost white hot so I threw on water which kept heat down. When boy came in I sent him up to put snow in the pipe, the fire immediately died out. Natives had never seen a chimney burnout before and supposed the roof was on fire. Cleaned out pipe and gave directions for future emergencies. Magazines got wet but not much else.[5]

It had been a close call, and they paid careful attention to the stove from that moment on.

During the first two weeks of December the weather was clear and windy, the barometer fell, and outside temperatures were -30°C while inside it was -9°C. Leffingwell spent much of it indoors. On December 8 Leffingwell received some mail, including a letter from Mikkelsen who wrote that Captain Leavitt had not landed much of their gear at

Point Barrow after all, and had simply thrown the dogs overboard to swim to shore.

Leffingwell looked forward to the arrival of winter and colder temperatures. Travel was easier when everything was frozen. It was a busy and enjoyable social time at the Flaxman base camp with many visitors. Toklumena delivered kamiks she had made for Leffingwell; he described them as beautiful with wolverine trim and "fancy" white tops, and a perfect fit. Mamayuok had also made him a pair of kamiks. These too had wolverine trim, vertical stripes, and patchwork at the top; the legs were made of caribou and sheepskin, and the soles of white seal. Mamayuok was also good at checkers; she beat Leffingwell six games in a row. Shugvichiak dropped by to give Leffingwell a lesson in the Inupiaq language. Leffingwell worked on the Okpilak River report and an outline of a more general report, using other U.S. Geological Survey reports as a template.

Leffingwell took pity on one of his favorite older dogs named "Dad." He wrote: "Let Dad indoors day & night, Poor old boy awfully thin. Give him a political job & by & by pension."[6] He packaged fossils to send to Dr. Weller at the University of Chicago, and organized his ornithological notes. Amidst the daily chores, visitors, and work on his report, Leffingwell read books about the Arctic and dreamt of other places, including Baffin Island in the Canadian Arctic.

By the third week of December, Leffingwell suffered from cabin fever, winter blues, and was not motivated to work on the report. When Grubin and Aigukuk arrived on December 20, he "forgot" to take observations to determine time. Toolik dropped by with a "beautifully beaded tobacco pouch" and Leffingwell wrote: "Wore my new fancy kamiks & feet very proud. Grubin beats me all to pieces checkers."[7]

On Christmas Day, he recorded the temperature at -28°. Everyone in the vicinity gathered at the house for an afternoon and evening celebration. They had a "shooting match" at a target, complete with a bull's eye in the center, followed by a "hula-hula," and finally an exchange of gifts. They dined on pemmican and vegetables. Leffingwell gave a plug of tobacco to each of the 10 men and a 10 pound tin of potatoes to each of the five women. He gave Shugvichiak an additional gift—a 50 pound tin of pemmican. Grubin and Aigukuk left on December 28 and

Leffingwell printed photographic negatives. He finished the Okpilak report three days later, and began preparations for the next trip.

As the year came to a close, Leffingwell's cabin fever continued. He was cranky about everything and everyone, including himself. He wrote: "Not having observations to work out nor anything pressing, I have loafed & read a great deal of trash after having finished all the decent books long ago. The house cools off early in evening so I read in bed after 7-8-9 pm. Do not sleep quite as well as last year, but less exercise."[8] That night it was -30° outside and -4° inside the house.

A New Year

During the first two days of 1908, Leffingwell was especially busy. Perhaps writing his thoughts helped motivate him, because he shoveled out around the observatory, cut out a new tent, finished it the following day, and hauled two loads of ice back to the house. Again he experimented with his socks: "Trying 2 wool socks & grass on one foot with fur sole, other foot 1 sock wool & fur slipper."[9] He made bread, something that he did every few days, then visited a sick member of an Inupiaq family. Leffingwell felt the cold through his clothes on his return, even though he wore several layers, some of which were fur.

Shugvichiak dropped by later, stayed for supper, traded some fox skins for flour, and they had a good visit. One of the conversations concerned religion. Leffingwell was raised in a deeply religious family and his father was an Episcopalian priest. But he also had strong opinions about religion and Arctic survival. "Conversed with Sagi [Shugvichiak] and said the missionaries who told him he would go to hell if he wore labrets, would go there themselves for lieing [sic] to him. They also teach not to hunt on Sunday even if hungry & game in sight. Rot!"[10]

There was a new moon on January 6 and Leffingwell took several sets of astronomical observations. He wrote that his hands were not cold, even though he recorded temperatures at -34° and -31°. Leffingwell also read about "prime vertical work" and wished for an instrument that would provide highly accurate readings: to 0.1 second for time, 0.5 seconds for latitude, and 0.5 seconds for azimuth.

Visitors and social events continued. On January 8 Arey and Gallagher, and a fellow named Sweeney, arrived at noon with a harrowing tale of survival. The trio had tried to locate the pass between the Sagavanirktok

River and the Koyukuk River drainage without success. They were out for 70 days and close to starvation. Their daily ration was three spoons of flour and a little blubber until Gallagher managed to shoot four deer, which ensured their survival. When they reached Flaxman Island, they could not see the ship because of a gale and the high snow drifts. Stories were shared until midnight. Flavina and Okalishuk, who had gone hunting, left Herschel Island on December 27 and returned to Flaxman Island on January 11 with five fox skins and two lynx skins. Shugvichiak and his family visited the following day and Leffingwell continued his work on occultations. The temperature dropped to -40°.

Leffingwell celebrated his birthday on the evening of January 13 with a hot toddy and his last cigarette. Arey, Gallagher, and Sweeney departed for home on January 14 borrowing two of Leffingwell's dogs. The temperature inside the cabin was -1°, but outside it was -40°C. After visiting a sick Inupiaq Leffingwell confessed to his journal that he did not know how to help. On his return to the house, he walked into a headwind while the temperature was -43°C.

Leffingwell took occultations and prepared for the upcoming trip up the Hulahula River. He sewed the tent, sharpened a knife, put a new sight on his rifle, and sewed a "wind guard" on his snow shirt. He carefully monitored his provisions, keeping track of how much he consumed, whether it was kerosene for the lantern or butter and flour, the rate of consumption, and how much remained for a given amount of time. When he had a surplus, he gave it away, often to Shugvichiak. While he worked, Okalishuk and Mamayuok played cards, and the winner smeared soot on the face of the loser. Leffingwell enjoyed the games and laughter.

By January 21 he acknowledged cabin fever and his desire to be outside again. He had designed and planned a new two-story house at Flaxman Island. (It was never built.) He finished packing the next day and iced the sledge runners to reduce friction. He calculated that his sledge was loaded with 550 pounds of gear. The other sledge carried another 475 pounds of gear and would be driven by Okalishuk and Flavina. They departed early the next morning and soon found themselves in another gale. Leffingwell judged the winds at 50 miles per hour and the team stopped to camp. Overnight the sledge was buried by the snow. On January 27, they reached Arey's home and temperatures were in the negative 20s. The weather cleared soon after, and Leffingwell took angles on the peaks and observations for determining latitude and time. Leffingwell marked his triangulation station with a "big nail in ground" and wrote that he would place posts the following summer. They stayed seven nights at Arey's, then moved to a house further east for two nights. Leffingwell recorded the temperature at -18° and wrote: "Measured off dist to 2nd station ≤ [about] 1600 m. Light poor can't see target even 900 m so use back of rod & sight wood cross pieces. Toothache all last night & today. Better after supper, so got out guitar & sang. Feel good. E breeze clds, clear night warm."[11]

Up the Hulahula River

The following day they traveled, covering five miles in as many hours and establishing seven stations as part of triangulating the area. The party then left the coast and headed south about 10 miles up the Hulahula River before they established camp 4 on February 6. The temperature was -20° the next morning when Leffingwell set off with Flavina to map the area. He packed his plane table and established seven more

Lunch stop on the Hulahula River, 1908. *USGS Photographic Library, Leffingwell Collection*

stations in about 16 miles before returning to camp six and a half hours later. By then, weather had moved in, bringing intermittent snow and poor visibility.

Leffingwell established additional stations as they traveled upriver and marked them with whatever was handy, including stones. On February 10 after traveling 22 miles, they camped with an Inupiaq group in the foothills near the Sadlerochit Mountains. One of the men shot a caribou, which they ate with fresh sourdough bread made in a frying pan.

It was too windy to work the following day, so Leffingwell left camp at 8:00 a.m. and hiked to the top of a nearby hill. He noted that the landscape was covered with glaciers. He took photos and his final entry that day was "fine clear moonlight night." They shifted to a camp further up valley among some willows just near the mountains. The wood assured a welcomed campfire. Leffingwell collected and identified fossils, while others fished upstream. Fish were plentiful; they caught 50 the first day and 40 the next. After Leffingwell took more observations, they cached some of their food and headed deeper into the mountains.

Camp along the Hulahula River, 1908. *USGS Photographic Library, Leffingwell Collection*

It was foggy all day and their visibility was limited at times to a quarter of a mile.

Leffingwell often hiked when the weather interfered with his work. On one occasion he climbed several hundred feet up the east side of the Hulahula Valley to determine if he could see Mount Chamberlin, which Leffingwell had named after the University of Chicago geology department chairman.[12] He did not see the mountain, either due to the weather or topography, but he was able to look up the Okpilak River valley, took photos, and returned to camp.

Leffingwell established additional stations with the help of Okalishuk and Flavina. One day they established five stations before it snowed and limited their visibility. They continued upriver and established camp 11 at the forks in the river. The wind blew at 15 miles per hour, too much for him to work. Okalishuk and Flavina hunted for sheep and returned with three rams. He wrote that the extra meat allowed them to stay

Mount Chamberlin, 1992. *Janet Collins Photographic Collection*

longer, and calculated the remaining food supplies: ten pounds of flour, eight pounds of bacon, six pounds of butter, and some cornmeal.

In his report written much later, Leffingwell reviewed the types of food provisions and the challenges of storage. He knew that meat was easily obtainable and could be stored frozen in the cellar at Flaxman Island. His diet included caribou, ptarmigan, ground squirrels, fish, an assortment of ducks, bear, sheep, and seal. He quickly determined that the amount of exercise and the climate increased the desire and need for additional fats in his diet. He recommended large quantities of butter, bacon, and salted pork be included in provisions and stored in air tight tins. Leffingwell also relied heavily on staples of flour, sugar, oatmeal, cornmeal, and rice, which required careful protection from moisture. He found that dried milk, dried fruits, and dried onions were best when stored in airtight tins.

Leffingwell wrote that the lantern had used more kerosene than he had originally calculated. He was concerned about the diminishing supply, so camp 11 would be their final camp upriver. He wrote: "gave up head of river." The headwaters would have to be explored another time. Leffingwell established his last station, with a view of Mount Chamberlin, took some observations, and noted that snow was blowing on the mountain tops.

On February 27 Leffingwell cached butter, tablets, and meat before they started the return downriver. They struggled against a 35-mile-per-hour wind that capsized the sledge and wreaked havoc with the party and dogs, then camped with a group of Inupiat and ate frozen fish. The following day they reached a sledge they had cached when they went upriver, but found it buried in five feet of snow. That evening they had a large meal of meat, vegetables, and crackers. He wrote: "Finest day this year. -33° while cooking outside & everybody with children sledding down snow drifts & men & women going & feeding dogs etc. Hated to go in at dark 6pm. Night fine, Venus, Jupiter & aurora & calm. Temp down 10° in 2 hours after sun behind mts."[13]

The return trip was slow; a mile to a mile and one-half an hour, when they were lucky. They trudged along through drifts and snow to their knees in high winds and temperatures in the negative 20s to 30s. They temporarily left a sledge behind until they had established camp downriver. Leffingwell described the following day as "calm and beautiful."

They again left part of their gear and traveled the remaining 10 miles to Arey's cabin on Arey Island where Leffingwell spent the following six nights. Temperatures reached -43° and -44° and winds alternated between calm and gale. Leffingwell waited for Gallagher to return with some blubber to feed the dogs. The party then camped at the Sadlerochit River delta sand spit for two nights and Leffingwell wrote: "Feet cold & feel cold snow thru skins. Not so bad as always at FJ [Franz Josef] land. Frost around face, 2nd time this trip."[14]

Leffingwell had traveled by sledge up the Hulahula River drainage and not quite to the headwaters during the winter. Snow limited much of his work, but he described what he could on the river's source, length, topography, the extent of glaciation on the sides of the valley and to the north, glacial till, and glacial moraines in the area. He noted that the effects of glaciation were not as obvious as the Okpilak drainage and indicated "probably that the main body of ice came over the divide from the South."[15] It was possible, he thought, that there were smaller glaciers he had not observed due to the snow. He explored the upper reaches of the eastern fork of the Hulahula River, and surmised that it was probably close to the Jago River headwaters. Current maps indicate that he was correct in his assumption. Others had indicated to him that the south fork of the Hulahula River offered access to a pass and route to the south. That is also correct. Based on the width and former size of the south fork stream, he also suggested that the eastern fork used to flow south, and had been captured by the Hulahula. He described future probabilities of stream capturing at several locations and identified thrust faults in the area. There is a narrow canyon where the Hulahula leaves the mountains and flows north. Leffingwell wrote that he thought it not possible to travel the area by boat or on foot. He had every intention of returning during the summer but would not.

It was March 14, late afternoon, and the wind blew 15 to 20 miles per hour as they finally reached the Flaxman house. They had been away since January 23.

A Brief Stay at the House

During the last two weeks of March, the house at Flaxman was once again the center of social activity as Inupiaq families dropped by to visit. Some brought fox skins to trade; others picked up items from Leffingwell to be mended or repaired and traded for food.

Leffingwell worked on the observations he had taken while at Arey's camp, developed photographic negatives, and made prints. He heated up a developing bath to process the negatives. It was a good time to be indoors. The weather was cold and outside temperatures dipped to -38°and -43.5°C. The inside temperature on March 19 was 1.5° C. Leffingwell wrote: "Ice all around bottom of walls, even frost on ceiling during this cold snap."[16] The weather was the usual mix of wind, overcast, snow, and clear, but the days were getting longer.

Up the Kugruak (Canning) River

Leffingwell agreed to pay Sweeney (the man who had been lost with Arey and Gallagher) $35 per month to assist him at the house and in the field. From March 31 to April 12, accompanied by Sweeney and Okalishuk, three dogs, and a sledge, Leffingwell traveled up the Canning to retrieve some sheepskins he had cached on the earlier trip. They took enough food for three weeks and extra to cache for the summer. The only scientific equipment he carried was a geology hammer and bags for specimens.

They traveled about 20 miles the first day. Leffingwell decided to cut to the river even though the others advised against it, and travel quickly became difficult. It was another lesson learned. They iced the runners of the sledge but the ice rapidly wore off and the snow was so soft that the sledge hardly moved. They shifted some of the load to their backs and left part of the load behind to retrieve later.

Shugvichiak decided to travel upriver too and was some distance ahead of Leffingwell. The next day he sent a sledge back to help with the load that Leffingwell had estimated at 650 pounds. They stopped every hour to ice the sledge runners. Temperatures were in the negative

30s. The party of three sledges reached the Ignek Creek and established camp 4. Camp 5 was then established four miles below Shublik Island on April 4. Leffingwell sent Okalishuk and Shugvichiak's son, Kissik, upriver to retrieve the sheepskins. Leffingwell looked for fossils and examined outcrops along the river while Shugvichiak hunted.

Leffingwell calculated the distance to the house on Flaxman at 55.5 miles. He was anxious to return home and start preparation for summer trips. And he was tired of not being completely mobile. He did not have any snow shoes and the snow was deep. It snowed as they began their journey back downriver. On April 9 and the following day near Shublik Springs, Leffingwell observed and described through drawings and his journal a phenomenon known as parhelia, sometimes called sun dogs. A parhelia occurs under calm, clear weather when light hits ice crystals in the atmosphere. The visual effect is a semi-circle or "ice halo" of sometimes vibrant colors curving down to the horizon.

Leffingwell arrived back at the Flaxman house at 2:45 p.m. on April 12 and quickly made tea as it was -20° inside. "Had it up to 0° before an hour...not tired but legs & ankles stiff & sore from running in soft & uneven snow."[17]

Icing the runner of a sled. *USGS Photographic Library, Leffingwell Collection*

At the House on Flaxman Island

Leffingwell continued to learn invaluable lessons from Shugvichiak and other Inupiat about snow conditions and icing sledges. He wrote: "Learned to ice runner with mouth full of snow water & hands. Lasts about an hour & sled pulls ≤ [about] 40% easier. Running with mouth open & eating snow yesterday set tooth aching again, not severely."[18]

Two days later, Leffingwell and Okalishuk loaded up the sledge with 325 pounds of gear and left for the mountains to retrieve the sheepskins. They traveled for five miles inland, then he sent Okalishuk to travel alone to the bluff while he returned to Flaxman. For the next two weeks, Leffingwell was intermittently sick. Prior to leaving he had lanced an infected tooth so it could drain and he could sleep. But something else was amiss. He was not sure whether it was food poisoning or flu, but he slept and read most of the time. When he felt better, Leffingwell cleaned the stovepipe, shoveled a ditch along the east side of the house for drainage, took photographs, and developed negatives. Sweeney chopped ice, and removed the snow and moss from the roof. It started to warm up outside but snow fell occasionally.

It was not until April 24 that Leffingwell tackled any tasks. He reviewed his lists of provisions for the next trip, a return up the Canning River. Arey and Gallagher arrived on April 27 after a hunting trip. Daylight hours were steadily increasing and by the end of April it was still light at midnight. Leffingwell hoped Okalishuk would return soon as he was tired of the house and planned to leave again as soon as the dogs were rested.

On May 2 Arey and Gallagher left, and Okalishuk returned during the evening of the following day. Leffingwell packed food for the next trip and wrote in his journal that there was a six degree temperature difference from inside to outside. They were ready to leave once the weather turned for the better. Leffingwell had observed and documented the geology up the Hulahula and Canning Rivers, but he was focused on spending more time up the Canning River region. Although he was looking forward to the trip and exploring the geography and geology, he was not looking forward to the struggle of traveling in 25 miles of soft snow.

CHAPTER 7

Return to the Canning

On May 6, 1908, Leffingwell and Okalishuk left Flaxman Island accompanied by Shugvichiak's son Kissik, two dogs, and a sledge weighing some 400 pounds. Kissik traveled with them to help with the sledge. As anticipated, the snow was soft and by the time they camped in late afternoon, Leffingwell's legs were sore and he was tired. The trip was easier for the Inupiat, who were in good shape and as Leffingwell noted "much younger."

In his detailed journal entries, Leffingwell recorded the photos he took, with camera settings and length of exposure, as well as detailed lists of birds he sighted, and the game they caught and ate.

Leffingwell's geology equipment included a rock hammer and belt, specimen bags, a gear bag, labels, rubber bands, indelible pencils, soft pencils, notebook and case, tissue paper to wrap the specimens, compass, large notebook, field glasses, pocket tape, and hydrochloric acid to determine the composition of rocks. If the rock "fizzed" when the acid was applied, it was a carbonate rock, likely limestone. Surveying gear included a small plane table, alidade, 20 sheets of paper, ruler, pencils, erasers, an aneroid barometer, tripod, triangle, scale, protractor, and thumbtacks.

The men cached food at the bluff and traded cornmeal for fish and ptarmigan with Inupiat who were in the area hunting. Five days later they reached the mouth of Ignek Creek where it flows into the Canning River from the east. Their route was not a straight line, but zigzagged overland adding distance to their travel each day.

The temperature warmed up to -4°C at noon and they saw fog up the valley. Ignek Creek was still frozen in most places. They camped on a sand bar unsheltered from the wind about six miles inland from the confluence with the Canning River. While Okalishuk hunted game, Leffingwell explored ahead to look at terrain. The snow was deep, the wind blew, and he traveled less than a mile. It did not look any better beyond. The next morning, both hunted—Okalishuk for squirrels and Leffingwell for ptarmigan. In the afternoon they soaked up the sun for a couple of hours, quite different conditions than the previous day. Leffingwell wrote: "Getting well tanned, eyes white from goggles, Look like owl."[1]

They stayed nine nights at the camp. The weather was mixed as usual; clear and calm, snow, wind, cold, and more snow. Okalishuk continued to hunt for game while Leffingwell sought fossils. He found 30 good samples that he concluded were Ordovician (a geologic time period within the Paleozoic Era, roughly 450 to 500 million years ago). During the next few days, Leffingwell collected an additional 200 pounds of rock and fossils to be carried out by sledge at a later date.

A diminishing wood supply and rising water near their tent forced them to move camp to the north side of a high river bank. On May 21 they switched to traveling at night because the temperatures were cooler and firm snow or frozen ground made it easier to travel. Leffingwell climbed 2,000 feet and constructed a five-foot high signal that aided his mapping. He took angles and drew a sketch map. The following day he climbed 3,000 feet and noticed some small purple flowers that he identified as saxifrage. The snow was rapidly melting. He made plans to head further up valley when Okalishuk told him that he was homesick and wanted to return to his family. Leffingwell talked him into staying another day to retrieve some food he had cached at one of Shugvichiak's spring camps. It was not an easy trip. Leffingwell wrote: "Reached edge of willows 2:10am & had awful time crossing them. One leg up to hips & crust breaking with small snow shoe on other. Shoe caught under crust & have a hard time to get out. This every step for 1/4 mile until I was very tired & angry. Reached other side 2:50am & couldn't find campsite."[2]

They finally returned to their camp but rising water forced them to relocate to higher ground. Leffingwell provided Okalishuk with a blanket, one dog, and food for his return and for the next six weeks, Leffingwell was in the field solo. He covered over 60 miles before he returned to Flaxman Island. Okalishuk had been the primary provider of game while Leffingwell worked on the geology; now he would have to do both. For the next nine days he explored the area around his camp. On the first day, even with snowshoes, he sank to his knees. He was exhausted, not feeling well, and his legs were tired. He was eating four to five pounds of meat per day and wondered if it was too much. The previous year he had eaten a similar amount and felt the same way. He wrote: "Started home 4:30 & waded up to crotch across some snow patches. Then tried short cut & got onto bad talus. Flat slabs, heavy enough to crush me, tilting very close once or twice."[3]

There were days when the fog was so thick he could not accomplish much. When in camp he sorted gear, made a pack harness, and rested. As his conditioning improved, he recovered more quickly. On May 30 he wrote that the weather was clear and the snow crust was hard and supporting his weight on his snowshoes. He hiked 3,000 feet above camp to the top of a peak, took angles, set up a signal for triangulation, and found more fossils. By afternoon, the snow was soft, a snowshoe broke, and he took a fall. It had been a productive day and although he had not eaten for 15 hours, he was not particularly hungry. Heavy fog rolled in and he turned in about midnight.

Leffingwell took the following day off and felt he had completed as much work as was possible. He was also concerned that his snowshoes would break within an hour of travel. The snow was melting fast, the rivers and creeks were rising, and stream crossings were uncertain on the return trip. Two days later it rained, and he placed willows under his bed to elevate it off the wet ground.

After breakfast the following morning Leffingwell observed some nearby ptarmigan and wrote that the female colors were brown with very little white. He recorded details of the behavior of two male ptarmigan, who appeared to be competing for a female.

Leffingwell traveled two miles up a creek above Shublik Bar, where he looked at outcrops, found more fossils at Shublik Springs, and noted similarities to what he had found in the Okpilak drainage. He cracked rocks around the fire in the evenings, and played solitaire for hours.

On June 8 Leffingwell continued along the north side of Ikiakpaurak Valley and camped for four nights while he collected fossils, cracked rocks, took angles for mapping, set traps for squirrels, hunted ptarmigan, observed and took notes on birds. River crossings were dangerous. At Eagle Creek he found himself up his knees in swift current and had to cross from one river bar to the next. In places where there was a main channel (rather than a braided channel), he described the current as flowing at eight to ten miles per hour, the depth at three to four feet, and stream width about 50 feet. It was intimidating.

He placed another triangulation signal, explored the geology and then found a glacial moraine, headed back to camp. The river had come up, crossing was well over his knees, and it was all he could do to cross with forward motion. There was ice in the river and the frigid water must have been painful but Leffingwell did not mention it in his journal. He was back in camp by 4:00 p.m.

Mid-June temperatures were generally in the 50° Fahrenheit range during the day and 40s during the nights, but at times it dipped into the 30s. There was often fog, especially in the evening and early mornings. Leffingwell described the usual pattern as "clear to N & heavy to S" in the mountains.

Arey had a four- to five-year-old cabin near the front edge of the Brooks Range. Leffingwell stayed there for three nights and continued his work. Along a bluff he found a fossil he identified as a crinoid, a first for him. Back at the cabin, he sat in the sun and worked on a base map using information from his triangulation stations and angles. It was June 15, and Leffingwell had not had the best of luck with fishing (he later became a fine fisherman), but he was pretty good with a gun, so when he saw a one-pound fish in a pool near the stream bank, he shot it and ate it for breakfast. He supplemented the fish at lunch with duck eggs from a nest near camp.

Leffingwell then headed upriver with food to last 12 days for him and the dog, to be supplemented by hunting. He traveled about 12 miles over six hours, then camped. He spent three nights at the campsite exploring. Near camp he saw tracks left by bear, wolf, and caribou. One day the temperature reached 17°C at noon and he heard the ice cracking beneath him. One evening he wrote his geologic notes, which took eight hours; during another evening, he worked on the map.

He moved camp twice over the next few days, performed his usual chores, and on June 21 Leffingwell spent a lazy day soaking up the sunshine, appreciative of the sun after a long winter. The wind and weather interfered at times with his loafing, but that was to be expected. After a day of mapping he wrote: "Had to lie down behind large boulder and hold plane table down, while I got a few bearings. Hands so cold that I could not draw well. Back 4pm and went to sqrl [squirrel] trap. Gone.

Had a sqrl in it but something went away with trap...Not heavy enough & not stuck into ground firmly enough. Fox or owl could pull it down. Cooked cornmeal for dog...Wrote up notes...drew up plans for instruments etc for future work."[4]

The following day the wind prevented him from using the plane table, so he took some important compass bearings. When the wind slackened late that evening he started his return down river. He took photographs, established camp, and when the weather allowed, set off to look at the geology. He found a large brachiopod fossil, weighing about two to three pounds and measuring five inches across. Temperatures warmed and mosquitoes increased. It was 22°C at noon and 21°C at 6:00 p.m. Before he reached Arey's house, Leffingwell camped, bathed, and wrote he was not used to the cold water. Exploring the geology of the area he discovered beneath the limestone a new formation that he had missed earlier on his way upriver.

Leffingwell reached Arey's house the following evening accompanied by a swarm of mosquitoes. He built a smudge of dried moss that night to create smoke to discourage the mosquitoes, but the smoke bothered him more than the mosquitos. Leffingwell stayed at Arey's cabin for four nights and accomplished a good bit more work. One day he climbed 1,600 feet and took photos of the geologic structure in the mountains. Another day he explored a valley, found more fossils, and located a geologic fault.

Leffingwell's report—USGS Professional Paper 109—is significant because of his accuracy in describing the landscape and conditions that remain relevant today. The Canning was the largest of the rivers he investigated. Similar to other rivers on the North Slope, there are two forks, of which the east fork is significantly larger. He described both forks as "trough-shaped" and he made note of a huge aufeis field, created by overflowing springs that freeze, layer upon layer, near the confluence of the two forks. An aufeis field sometimes extends across an entire river making travel by boat or raft difficult or impossible. The extent of his travel directly on the Canning River was limited to a "short distance" up the east fork, but he did explore to the east of the river.

Leffingwell recorded the physiography from the confluence to the delta. He did not find drift evidence of ice in the Ignek valley or in the Ikiakpaurak valley. A distinctive feature on the west end of the Sadlerochit Mountains is "Red Hill," easily seen from the Canning River. Upon close examination he determined that during the glaciation period, ice had reached near to its summit and represented the limit or boundary of glaciation in that area. He remarked that Shublik Springs had cut through moraines and exposed many boulders before emptying into the Canning River. The western end of the physiographic Sadlerochit Mountains and the Shublik Mountains are just east of the Canning River, but the geologic structures extend further west. Leffingwell concluded that the Canning River had possibly flowed across the area between Red Hill and the west end of the Sadlerochit Mountains because he could not otherwise explain the gap.

Forks of the Canning River, with aufeis covering the flood plain. *USGS Photographic Library, Leffingwell Collection*

Ignek Valley with Sadlerochit Mountains on left. *Gil Mull Files*

Red Hill near the Canning River, 2008. *Janet Collins Photographic Collection*

Leffingwell wrote about the change in vegetation near and around Arey's house since his previous stay. Warmer temperatures brought sweet smelling flowers and long green grass. Within a day or two he also admitted to a "little longing" to return to the house on Flaxman Island. It was now late June, and he had been away since early May.

In a twenty-four hour period from June 27 to June 28, temperatures plummeted 16.5°C. He spent the day in camp, most of it inside the tent. When he awoke the next morning, the snow level had dropped to 1,000 feet. Leffingwell broke camp about noon and traveled 13 miles in under seven hours to reach one of his earlier camps and caches. He explored a nearby creek, located another outcrop, and gathered more fossils.

Leffingwell had planned to build a raft using the sledge and willows in order to transport his gear and rocks, but the aufeis was still four to six feet thick in the river and he was concerned that it would break off into the narrow channels and block the route. He decided to abandon the raft idea, packed 80 pounds of fossils down to Ignek Creek and cached them.

On July 1 when he took the temperature at 4:00 p.m., he recorded it as the warmest day yet, 24°C (75.2°F). He had suffered occasionally from an infected tooth. It bothered him again that day, so when he returned to camp, he: "shoved needle into gum. no knife needle too small, so used sail needle, blunt, but made in enough hole to drain gum. Ache stopped ok. read till 12:30 & turned in."[5] While at the camp, he cached the following: "Corn meal, 35 lbs, granola 8, pem tin, 10, butter 6, salt 3, ...2 tins, clothes bag, gun 70 ctgs, stove & pipe, sled & cover, tent cover torn, blanket, wool, blanket, rubber, fry pan, 2 tin plates, ...geol. rocks. 7+- dried vegetables."[6]

Leffingwell retrieved fossils over several days and decided to haul the loads to the bluff. He started with 100 pounds and loaded the dog with 50 pounds. It was not long before he cached 60 pounds of his load and 30 pounds of the dog's. Leffingwell described what he called "homeward bound fever." Even the mosquitoes did not deter him. He camped and when he looked out of the tent, wrote that he could barely see the dog for the mosquitoes. He brought the dog inside for the night and spent a fair amount of time eliminating mosquitoes that had accompanied the dog inside.

He continued his march for the coast on July 4 and found even worse conditions:

> Afterwards wind started up & shifted to NW & mosqs in millions. Cutting with wind, had to keep handkerchief waving before face all the time, stopping every few min. to clean them off the dogs face. Especially bad when crossing wet land to save going around meanders. From 3-7 stopped only 3 times to light pipe & get a drink. About a million on dog pack sack & on my back. At 7pm, started to sprinkle & stopped at a creek with few stunted willows. Built smudge & in 25 min. had tent up...Had him in tent all night last & in now.[7]

It had been a long day of walking over marshy wet tundra and tussocks, and the soles of his shoes were worn through. He camped at 10:00 p.m. and the mosquitoes hummed him to sleep.

As Leffingwell neared the coast, the breezes increased and the mosquitoes lessened. Along the way, he stopped and sketched the river, and assessed the feasibility of crossing. It was still too high so he made camp. He broke camp at 9:30 a.m. the following morning to locate an easier crossing along the east branch of the Canning. (Like most of the rivers on the North Slope, the Canning River is braided, its delta extending over ten miles from west to east.) Leffingwell thought he saw the house, but was not certain. He knew he could see the eastern end of Flaxman Island. He observed and recorded the birds he saw: skuas, plovers, arctic terns, and snow buntings. That evening he decided to leave the tent and gear behind and head for the coast, weather dependent. He planned to carry biscuits, chocolate, extra socks, matches, and exposed camera film. Crossing the river might be impossible so his backup plan was to travel to the coast and launch a boat, or wait for someone to come along that could provide transportation.

On July 7, he realized he could not cross the Canning River without getting totally soaked or worse. But he attempted the crossing anyway. He left the tent up with the gear inside, just in case he needed a quick and dry retreat, and left camp at 7:30 a.m. A cold wind was blowing about 15 miles per hour. He had lost his mittens so he used socks in their place and later penned this account:

> At 8:30 entered water (a few inches less than yesterday by gage at tent) when fog shut down, but not thick. Crossed a lot of water & came

> to place where I couldn't see across, so pulled out & followed down a couple of miles, but water gathered into one channel & couldn't cross it. About to give it up when fog lifted & I decided to take another look. Back to same place I left at 10:30 and picked out passage, not over knees except one place. With 2 sticks to balance with I tried it, and nearly had a ducking. Water carried me along into deep place and I was wondering how soon I should have to swim, when it threw me against a shoal place & I was across E branch. Thought I was in up to arms, but high water mark only came to 4 inches below belt, as it was, I was helpless & only hitting the high places for some distance. Water not so cold, but wind cut when I got out. Had one more deep place, but not much current. Made for higher land (6 ft) between rivers & had some more water, but shallow. Reached bank at 11:30 = 3 hours in getting across & one hour straight ahead. Built fire & wrung out socks & senne grass & warmed up. If I had lost footing & got wet all over, I should have had a cold time of it. Matches & paper in tight milk tin, but couldn't get dry on a foggy & windy day.[8]

Leffingwell later crossed over another channel; the water was up to his hips, but it was easier because the bottom had a hard surface. He was pleasantly surprised when he reached Brownlow Point and discovered Inupiat camped there. In his absence, one of the women had made him a new attege and fur breeches. He was grateful, changed clothes, and ate salmon for dinner. Later in the evening four young Inupait rowed him across to Flaxman Island.

Flaxman Island

When he arrived at the house at 10:00 p.m. on July 7, temperatures were about 3°C. Leffingwell had traveled 33 miles that day and his feet were understandably sore. He turned in at midnight.

Upon his return, neighbors dropped by and they enjoyed a "big feed." All of the Inupiaq families had relocated to within one-half mile of the house. His trade with the Inupiat had steadily increased since his arrival in 1906, and he hoped to pay for some of his expenses by selling the animal skins he acquired. Leffingwell cleaned, organized, and stored gear and skins. He also went for walks and on one occasion discovered that an eider had nested behind the house. Snow buntings were also plentiful.

On July 8 he observed that the ice was still close to the shore. One afternoon he looked at fossils in his geology books. Another day he traveled about a mile to the west end of the island to scout a new location for a house, noting that the current one was "wet and full of holes." To the north of his house, one of the Inupiaq families was building, and another family was considering building quite close to Leffingwell's place. It is possible that Leffingwell was uncomfortable with the idea of close proximity to anyone. Although he very much enjoyed the company of others, he was accustomed to living alone. Leffingwell stayed about a week before he headed upriver to retrieve his gear from his last campsite.

Retrieving Fossils

On July 15, accompanied by Okalishuk, Toxiak, and another Inupiaq, Leffingwell started for the Canning River to retrieve the load of gear and rocks he had cached. They borrowed an oomiak from Okalishuk's father and traveled close to the shore of Flaxman Island before paddling across to the mouth of the river. Although the boat carried a light load and drew only five inches, they became stuck a few times on the mud flats. The Inupiat and dogs lined the boat up braided channels while Leffingwell paced off an area between triangulation stations. He carried a small plane table for mapping and used wood to mark the stations. He found that he could see the markers from "2,000 paces." The group traveled 12 miles in seven hours before camping; when Leffingwell finally turned in, he wrote how pleasant it was to experience wind and rain after being indoors for a week.

On July 17, in mid-morning, they reached the tent he had left erected and the cache on the bluff. They had lunch and Leffingwell hiked off toward a glacial moraine that he had decided to use for his triangulation network. He constructed a three-foot rock cairn and a tripod of sticks on which he placed white cloth, hoping it would be visible from the house. He also took sets of angles in eight directions, as they would be critical in establishing an accurate triangulation network.

That evening Okalishuk headed for the cache of rocks and the other two Inupiat left to hunt near the Hulahula River. Leffingwell's work was limited by the wind and hoards of mosquitoes during the next two days.

Using the plane table required calm weather, but the mosquitoes drove him back to camp and his tent for the day, where he played many games of solitaire before turning in at 9:00 p.m.

It was warm and raining two days later, so he spent another day tent-bound, playing solitaire. The mosquitoes swarmed into the tent at any opportunity. He was worried about Okalishuk who had not yet returned, and was concerned that if the river level dropped too much, the boat would be grounded.

Finally, on the seventh day at the camp, Okalishuk reappeared.[9] Leffingwell was angry with him until he learned what had happened. One of the dogs had run away and Okalishuk had looked for him for several days. In order to return to Leffingwell as quickly as possible, Okalishuk had traveled 25 miles without sleeping.

Leffingwell and Okalishuk tried to cross the river the following day, and Leffingwell fell in to his waist. He emptied his boots and decided to repair the badly leaking boat. There was one hole the width of three fingers, so he simply sewed up the canvas. They ate lunch and decided to camp that evening near some pingos (dome-shaped hills consisting of an ice core, formed by water under pressure that pushes upwards as the ice core expands) and the mouth of the Canning River.

Winds usually came from the east so Leffingwell oriented the tent accordingly. The next morning it blew 25 miles per hour, but from the west. He worried that the tent cover would blow off, and it did. The wind calmed that evening, the missing dog, now "tired and fat," rejoined them at their new camp. Leffingwell and Okalishuk headed for home under rainy skies. They arrived back at the Flaxman base camp late in the evening on July 26, and Leffingwell wrote that the house felt damp from the rain.

Back on Flaxman Island

Leffingwell successfully collected significant geographical and geological information about areas that were previously undocumented. His accomplishments to date had included the pack ice trip with Mikkelsen and Storkerson, and trips up the Okpilak, Hulahula, and Canning Rivers for mapping of the geography and geology. He established the required triangulation stations in those areas for accurate mapping, collected fossils and other rock samples, and wrote an article about the

"Flaxman Formation." And he spent a significant amount of time working on the Flaxman Island house. Leffingwell determined that he did not have adequate equipment to continue to accurately map the geography and geology. It was time to return to the lower 48.

He planned to return the following year to continue mapping. He began carefully watching the movement of the pack ice as he prepared for his departure, planning to board one of the passing whaling ships. Whaling ships served an important role along the north coasts of Alaska, providing local transportation, provisions in exchange for animal skins, mail service, and visitors. On July 27, he noted that the ice had moved offshore, although it not enough for a ship to pass. The temperature ranged from 2°C to 8°C (35°F–46°F). The winds blew at 50 miles per hour on July 29, shaking the canvas roof on the house. The next day the pack ice was one to two miles offshore, enough for a ship to pass. Leffingwell immediately packed rock specimens in boxes and stenciled the addresses on the outside. He took further observations, calculated earlier observations, took apart instruments for cleaning, and organized the outside rack for the provisions that would arrive with the ships.

Arey's family arrived on July 31, as did Grubin and Eskal. They brought mail for Leffingwell and set up tents outside the house. One piece of mail was from the Royal Canadian Mounted Police at Herschel Island. Leffingwell had written the RCMP in March about proper compensation for Flaveenek [Flavina] and Mamayuok. Originally they were to work for Leffingwell for two years, but they were with him from October 1907 until March 1908—only about five months. The RCMP wrote that Leffingwell had paid Flaveenek too little, and that Mikkelsen had promised Flaveenek the bear skin he shot. Fairness and reputation were important to Leffingwell, so he supplied an Inupiaq who was traveling east with a can of kerosene in exchange for transporting some items including a bear skin for Flaveenek, and six fox skins. The fox skins were payment to the RCMP for a rifle they had provided to Mikkelsen.

Leffingwell turned his attention to the movement of the pack ice and concluded that the ships would soon arrive. The weather had also improved. By August 3, more Inupiat arrived at Flaxman Island for trading, totaling 15 tents. Caribou skins were at a premium and a trader named Wixra had 60 of them. Leffingwell had far fewer.

Over the following three days, six Inupiaq boats left Flaxman headed west. Leffingwell was concerned about his diminishing food supply, and had promised most of his remaining flour to Arey and Grubin. He was uncertain if the whaling ships would stop at Flaxman Island to pick him up. Leffingwell wrote: "Up late & loafed & read all day. Can't work & can't go away for fear ships might show up. Feel that I am wasting time but can't settle down to do anything. Taking looks with telescope for ships & getting anxious. May be in such a hurry that they will not stop as last year. Have about a pound or so of flour & cornmeal is mouldy."[10]

As the days passed, Leffingwell read, made observations for time, weather, and ice, and tried to get the sheep skins dried out. He worked on maps, copying sketches, reducing scales, and working on map projections. On August 9, the ice was blown closer to shore, making ship passage impossible. Five days later it was foggy, rainy, windy, and the roof was leaking. Leffingwell was miserable, anxious, and ready to leave. There were no ships on the horizon, and winter was fast approaching.

Return South

On August 15, 1908, Leffingwell speculated with Arey and Grubin about whether the ships would soon come to take them to Barrow, and, more importantly for Leffingwell, south to the lower 48.[11] They decided to leave for Barrow by small boat if none of the whaling ships arrived by August 20. The ships had always been good about stopping at Flaxman Island, but his concern was growing daily.

Over the next couple of days, Leffingwell continued preparations for departure, took soundings, measured angles, and continued mapping. He read all day Sunday, August 16, and noted in his journal that the ice was closing in once again.

The following morning, Leffingwell was awakened at 7:00 a.m. by an Inupiaq who sighted a whaler—the *Karluk*—arriving from the east. Leffingwell hurried to hail the ship. He was surprised to find Stefansson on board heading to Barrow to obtain food provisions. Stefansson was irritated with Leffingwell and later wrote: "I made here another attempt to get matches, but although Mr. Leffingwell had some he did not consider he could let me have any without breaking faith with certain Eskimo among whom he had promised to divide them."[12]

Travel conditions were not good. The *Karluk* followed a careful path from Flaxman through the ice before anchoring at Cross Island. They finally reached Point Barrow after two days of rain, fog, and snow. Anchored to an old ice floe that evening, Leffingwell noted that the route west of Barrow "looked solid" with ice. At Point Barrow, he and Stefansson traveled to shore in a whale boat, and arrived at midnight. Leffingwell camped on the beach and crawled into his sleeping bag about 4:00 a.m.

The following morning, Leffingwell and Stefansson walked up to Brower's trading store at Barrow. Brower invited them to stay at his place while they waited for a ship to take them south. A week later five ships arrived around noon. One was the *Narwhal,* which landed a boat on the beach to pick up Leffingwell, and delivered his mail and freight. On board the *Narwhal* was Storkerson, one of the original AAPE members. Leffingwell advanced Storkerson $250 in wages to come back to work for him the following summer.

For two days Leffingwell organized and stored provisions with Charles Brower and Tom Gordon, for use after his return north the following year. The provisions included groceries, clothing, tool chest, photo supplies, telescope, geologic notebooks, boards, sheet iron, and a tarp. Leffingwell sent 25 pounds of flour to Shugvichiak on Flaxman Island along with other items.

The first week of September passed aboard the *Narwhal.* Whaling ships were still anchored near Point Barrow, taking whales. The temperatures dropped into the teens, and it snowed. Leffingwell wrote: "On board Narwhal...Saw ships coming am & got gear down to beach. On board pm & have fine bunk in chart room. Desk, etc, etc."[13]

Shortly thereafter, the *Narwhal* left Point Barrow and headed west. They followed the fluctuating edge of the pack ice, traveling north at times or even northeast and anchoring at night. Other whaling ships often traveled in close proximity and on one occasion the *Jeanette* provided the *Narwhal* with several "boat loads of fresh water."

Leffingwell worked on his mapping and calculations, until he was hit by seasickness. When the seas were rough, he skipped meals and lay in his bunk. On calm days, Leffingwell went outside on deck and observed "jelly fish by thousands" and took notes about the birds he observed: phalarope, ivory gulls, long-tailed ducks, and what he thought were black guillemots.

On September 21 the temperature was -12°F and the seas were starting to calm after another gale. It was a damp cold that chilled to the bone and it was snowing. The *Narwhal* was east of Herald Island, off the Siberian coast by September 24 and the following day Leffingwell started working on his report. The *Narwhal* was headed northeast once again and on September 26, lowered boats to continue their hunt for whales. They took one and Leffingwell noted: "9 ft bone & about 1600 lbs" and according to him worth about $8,000. Twice, Leffingwell went aboard the smaller boats as they unsuccessfully attempted to capture a whale. The first time they were out for nine hours, and spotted about twelve whales. Several times the whales came quite close—fewer than one hundred yards. The second time they were out for about four hours and it snowed so hard they lost sight of the ship for a time. Temperatures ranged from 18°F to 30°F.

By October 1, the weather and seas were calm and Leffingwell wrote that a whale had come within fifty feet of the *Narwhal.* Captains from nearby whaling ships gathered on the *Narwhal* the evening of October 4. Their consensus was that the "whales have been acting queerly," likely because they were venturing so close to the ships. On October 12, they sailed south through the Bering Strait. Seas were rough, winds were strong, and the temperatures in the 30s. Leffingwell was seasick and confined to his bunk. When the seas calmed, he reported having breakfast and "several suppers." Near Cape Prince of Wales in the Bering Strait, he observed about 1,000 walrus sleeping on the ice. They also saw several killer whales en route to the village of Unalaska, arriving on October 19. Leffingwell wrote: "Nearly calm inside, fine smell of hills. Warm in flannel shirt. *Thetis* here." His field journal ended with that entry.[14]

The many pages that followed were full of notes, lists, and drawings. From conversations aboard the *Narwhal,* Leffingwell made notes for future reference, some about companies in the lower 48 that supplied various goods to Alaska. As he mulled over his fieldwork, he made a list of questions including one about the "accuracy with which triangulation can be carried on by Lat [latitude] & az [azimuth] on peaks without signals."[15] He was not sure if his work would be as accurate as he hoped.

His detailed notes included bird observations, caches of food, planned projects for the Flaxman Island house, food lists for trips, items to purchase for his next trip north, items for Arey, tasks for Storkerson, a plan for work over the next five years, and some phonetic translations of Inupiaq words. He included calculations for map scales, generated a comparison chart for Celsius and Fahrenheit, and based on an earlier conversation with Arey wrote notes on how to tan hides. Hoping to generate some income from trapping and trading, he kept close track of the prices paid for pelts of different animals.

His drawings also included plans from various perspectives for the house he intended to build in 1908–1909. The new house would be 24 feet by 16 feet or 24 feet by 14 feet, and he detailed the layout inside. He also drew the traditional dome tent with the stovepipe. He depicted the tent vertically and laterally, and calculated the amount of space based on the different dimensions. His other drawings included a canoe and dimensions for Arey's oomiak.

Leffingwell drew several primitive and generalized maps, one that showed the area from the eastern Sadlerochit Mountains south to Schrader and Peters Lakes. He identified major drainages and some physiographic features. He drew another pictorial map that extended from west of the Sadlerochit River to the Kongakut River near the Canadian border. And he drew a general map of the coastline that included the location of Arey's cabin.

Soon enough he would obtain better instruments, tools, and more provisions for his return north in 1909. But for now, he was on his way home to family and friends after being away since early 1906.

CHAPTER 8

Return North, 1909

When Leffingwell departed northeastern Alaska in fall 1908, he planned to return in spring with higher quality scientific equipment to continue fieldwork and mapping. But: "The whaling ships, which were the usual method of transportation were not going north that year, so I got a gasoline yawl in Seattle, a thirteen-ton craft. The Chicago Yacht club had made me an honorary member, so I sailed under their colors."[1]

Leffingwell did not elaborate on why the whaling ships were not traveling north; presumably ice conditions prevented them. Rather than wait, Leffingwell decided to contract the construction of the yawl in Seattle and named it the *Argo*.

His parents provided the funds from Leffingwell's future inheritance. The 50-foot yawl had a 10-horsepower engine that made four to five knots an hour, and the yawl drew less than four feet. He hired three crew members for the journey: a mate, a first mate, and a steward. The trip and the crew would prove among the most challenging of his experiences in the north.

Leffingwell was finally ready to depart Seattle for Flaxman Island and the North Slope of Alaska on May 20, 1909. He was up at 5:00 a.m. and found that two of his three crew were not aboard, after spending the previous evening onshore drinking. He managed to get them aboard by 11:00 a.m. and they shoved off less than two hours later. The vessel's builder accompanied them a short distance to take photographs. As the crew slept off their hangovers, Leffingwell worked the yawl alone. The weather was relatively calm, but they were bucking a strong current and some wind. It was slow going, eight miles in three hours.

Prior to sailing Leffingwell started a Pilot House Log Book to maintain details of the trip. He was, as always, meticulous about details and would record the time of day when the anchor was weighed or dropped and the location, the distance traveled and speed, the bridge compass heading, and occasional notes on the sails used and water depth. He also assessed the workmanship of the *Argo.* Prior to departure Leffingwell expressed concern to the builder about the proximity of the engine exhaust and the galley stove to the woodwork and the builder assured him that all was well.

Sure enough, within several hours of departure the woodwork started smoking from the heat of the exhaust. Leffingwell stopped the engine for a half hour while he threw water on it to cool it down. The builder had indeed taken shortcuts. Leffingwell later acknowledged that he knew he should have turned back, but time was precious and the window for reaching Flaxman Island was narrow. Leffingwell continued to throw water on the hot engine at intervals and tried to protect the wood with pieces of tin. As the woodwork continued to smoke, he worried about favorable winds and making as quick a trip as possible. The ship anchored in Port Townsend, Washington, at 9:00 p.m. Since their departure that morning from Seattle, Leffingwell had monitored fuel consumption and engine efficiency. It was not surprising given his meticulous nature and attention to detail. And it was a new vessel. He was curious to see how it would perform.

Leffingwell wrote of his "deep-seated dislike" of sea travel, and yet he was determined to return north. If navigating his own yawl without experience was the solution, so be it. "My own qualifications were limited. I had never navigated a vessel. But I did have a good grounding in astronomy and I hoped by that means to lay our course safely...We were not going north by the Inner Passage, and I wanted a strong offshore wind to blow us far from land and leave little likelihood that we'd be blown back again."[2]

Saving time meant sailing in a direct line from Vancouver Island to Unimak Pass east of Dutch Harbor along the Aleutian chain of islands, rather than the more protected route, the Inside Passage, which follows the coast of British Columbia and southeastern Alaska. Achieving Flaxman Island was limited by a narrow window of opportunity and open water between the northern Alaska coastline and the ice pack.

Departing Port Townsend into the Strait of Juan de Fuca, the *Argo* encountered strong west winds that forced them to return to anchor that afternoon. Leffingwell's dangerous assumption that the builder had provided quality workmanship in the construction of the yawl proved incorrect time and again during his travels north. Leffingwell noted that the chain provided by the boat builder was about 18 to 20 fathoms long instead of the usual 40 fathoms. Anchoring would be impossible at various locations throughout the voyage.

It took the *Argo* a week to travel from Seattle through the Strait of Juan de Fuca. They successfully anchored in Port Townsend, Port Angeles, Crescent Bay, Clallam Bay, and Neah Bay along the way. As they headed out into the open ocean past Cape Flattery at the northwestern tip of the Olympic Peninsula in Washington State, the swells were heavy and they were using both sails and engine. It did not take long for Leffingwell to become miserably seasick. There were ongoing issues with engine performance. The overheating galley stove caused the wood behind the stove to smoke whenever they used it. They continually threw water on the wood. On May 27, Leffingwell wrote: "Rain most all day, wet, cold & vomiting & thoroughly miserable. Got a fire & had warm meal. Poured water behind stove. Nearly calm all pm & night."[3]

On May 30, conditions worsened and he described their plight: "waves breaking too hard so heave to under mizzen & staysail."[4] Leffingwell was using two sails to counter each other in an attempt to slow or stop the boat. Otherwise, there was a potential danger of traveling too fast and out of control. Two days later, he described the seas as "worst ever for me." The barometer was fluctuating and the winds constantly changed direction. Sailing conditions were treacherous and, at one point, Leffingwell placed oil bags over the side of the yawl to keep the waves from breaking over the boat. The *Argo* was being blown off course to the northeast.

By June 7 Leffingwell was still miserably seasick, there were heavy swells, waves crashing over the sides, and winds occasionally in excess of 50 miles per hour. The distance traveled each day varied widely, and they covered anywhere between 50 and 100 miles and at an average speed of four and one-half knots per hour.[5]

On June 12, 24 days after leaving Seattle, a dispute among the crew flared up. Punches were thrown and Leffingwell "got them to call it off for a while." The challenging journey and related stresses did not help and certainly contributed to flaring tempers. The men must have been exhausted. They used the engine when they could and put out the oil bags for stability when the swells threatened their safety. On Saturday,

June 19, Leffingwell wrote: "Only 5 miles from 10pm to 7am. light breeze...Heavy swells knocking us to pieces. Barom high 75.3 rain & sun. Can't use engine, nor sail against SW swell. All hands discouraged, stuck 4 days same place. Can't sail against one swell...gale after gale, can't even run with fair wind."[6]

Dutch Harbor

They sighted land on Monday afternoon of June 21, after a month of sailing in some of the worst conditions Leffingwell could have imagined. They couldn't anchor until the next day in Unalaska harbor. Leffingwell was exhausted. He wrote: "*Karluk* here. Had less than four hours sleep since Sun[day] and bad head from gasolene & very nervous. Capt. Cottle sent over boat for me & I went over to brkfst, but was too sick to talk much."[7] The trip, one of the most dangerous Leffingwell would ever endure, did not change his dislike for being at sea.

Dutch Harbor provided another opportunity for two of the crew, James Boyle and James Clark, to get drunk. One of them started throwing items overboard and Leffingwell had to restrain him. While Leffingwell was ashore the next day, the most troublesome crew member did a bit of work on the *Argo,* then helped himself to some of Leffingwell's alcohol. When the man again became drunk the following day, Leffingwell only reprimanded him. He needed his help, such as it was, for a while longer.

Leffingwell had purchased a small whaling boat from the whaling ship *Narwhal* for use along the north coast. He took delivery of it while in Dutch Harbor. Captain Cottle of the *Karluk* offered to tow the *Argo* north, and secured the small whaling boat aboard his ship. Leffingwell wrote: "Feel much relieved. Very much wrought up lately with complications."[8]

His anxiety only increased when he discovered that the wood beneath the stove was charred through. He continued: "This is awful to think of." The *Argo* could have easily sunk as a result, and likely would have on their journey north if he had not discovered the charring. Leffingwell solved the problem by moving the stove away from the wall, placing a three inch pad of concrete beneath it, and adding sheet iron and asbestos around it.

They departed Saturday, June 26, at 4:00 a.m. from Dutch Harbor after loading water and coal supplies. The *Karluk* towed them at six knots and by noon had traveled 40 miles. The following day they traveled 169 miles using the engine and then later, only the sails. Given his concerns about the narrow window of opportunity to reach Flaxman Island before the ice closed in, Leffingwell must have felt optimistic about their progress.

The south-southwest winds picked up to 20 to 25 miles per hour on Monday June 28, so they switched to sails. Still under tow the Argo at times was pushed dangerously close to the *Karluk*. At 9:30 p.m. that evening the tow line broke and the Argo crew watched the *Karluk* disappear into fog. Leffingwell estimated that they had covered 117 miles that day. They made 110 miles the following day despite the fog. The *Argo's* fog horn sounded from time to time as warning to other vessels. They heard a fog horn behind them and were eventually passed by a revenue cutter steaming north although Leffingwell did not or perhaps could not identify which revenue cutter. The following day they traveled only 19 miles.

They arrived in Nome shortly after 11:00 p.m. on July 3, and stayed a week awaiting further funds from Leffingwell's parents. They anchored in the company of seven ships: a couple of whaling ships (including the *Karluk*), revenue cutters, and a schooner, the *Mary Sachs*.[9] While in Nome, Leffingwell located Sweeney (the man who had been lost for 70 days with Arey and Gallagher in 1908) and convinced him to join the crew of the *Argo*.

The tides were low Saturday morning when they tried to leave, and the *Argo* grounded. Even kedging the boat did not help. They finally deposited three tons of provisions on the beach and with a line and some help from another ship were able to get the *Argo* afloat in deeper water. By 1:30 p.m. they had reloaded the *Argo* and were headed to Sledge Island from Nome. They anchored about 60 yards off shore in rough winds and seas. The crew was still recovering from their alcohol binge.

They had a layover day because of the continuing gale, and Leffingwell was feeling depressed. He had spent 10 days and a significant amount of money in Nome without anything to show for it except a drunken crew. Their journey continued at 2:30 a.m. the following morning under sail and motor. By noon they were traveling at six and one-half knots, helped along by a southwest and southern breeze.

They set anchor at 4:00 p.m. near Teller (around 100 miles from Nome) on Wednesday, July 14. The two crew members promised not to get drunk if allowed to go to shore, but despite past experience, Leffingwell let them. They had not returned to the *Argo* by 11:00 a.m. the following morning, so Leffingwell had the dinghy retrieved from shore, stranding the two. The following morning Leffingwell acquired more water for the voyage and talked Sweeney, one of the two crew members left ashore, into traveling with them to Barrow. He felt that he, Joe, Sweeney, and Clark could handle the *Argo* for the remainder of the journey to Flaxman Island. To his credit, Leffingwell gave friends of the crew member left behind his "last $20.00 for food for him when he sobers."[10] Leffingwell had no regrets and likely wished that he had left the drinker behind earlier.

At Point Spencer on July 16 they met over 40 Inpuiat in two large canoes and traded sugar and tea for ducks and dried salmon. The *Thetis* also arrived and anchored nearby. The weather calmed so they hoisted anchor and headed north for Cape Prince of Wales about 7:00 p.m. on July 18. The engine's check valve was malfunctioning and required monitoring.

From Cape Prince of Wales to Point Hope was a distance of 193 miles. When they anchored near Point Hope two days later, Leffingwell calculated that they had traveled 113 miles in a 24-hour period, the longest consecutive distance yet. Thirty-two Inupiat in three canoes, along with a missionary and teacher, visited the *Argo*. Leffingwell had supper later that evening with the missionary and learned that he had met Mikkelsen in the Yukon in 1907. A gale came up on July 22 so they sailed into a lagoon 12 miles east of Point Hope and stayed until July 26. Leffingwell spent the extra time reading. He had finally started to relax and wrote: "Most pleasant days since starting. In harbor and can't go on & no anxiety about anything."[11]

They carefully worked their way north through the ice and ended up hugging the shoreline as they went. They anchored near the village of Icy Cape, where Leffingwell traded for seal hides. The following afternoon winds were blowing out of the north-northeast, forcing them to anchor near Point Belcher. Leffingwell wrote that the pack ice was probably close in to shore ahead of them. Sure enough, he saw the whaling ship *Herman* in heavy ice the following day, and went aboard for a brief visit. Captain Bodfish gave him mail and funds from the British explorer Harrison to pay a debt to an Inupiaq on Herschel Island. The *Argo* continued through ice into open water and by 8:00 p.m. had tied up beside the *Karluk*. Leffingwell passed the items along to Captain Cottle of the *Karluk* later that evening.

Both ships heaved anchors shortly after noon the following day to continue north. By 3:00 p.m. August 3, the *Argo* anchored about 20 miles southwest of Barrow near a creek and a group of reindeer herders. Leffingwell wrote: "Ice immediately ahead shoved up on beach & no path for us. Canoe load on board. Learn that Konalooa is camped only 3 miles away. He ought to be at Flax Id. hunting for me. Ashore & had a bite with natives."[12] As usual, conditions quickly changed. By 8:00 p.m. the ice was moving out, but the barometer was falling and it was raining hard. By 9:30 p.m. there was a clear path free of ice and the rain had lessened, so they shoved off. Within an hour they encountered thick fog, headed close to shore, and tied up to a grounded ice floe.

Leffingwell slept most of August 7, then went for a walk on the ice, took photographs of the *Argo,* and wrote: "All fire wood gone...have to saw up 2 x 4's for house now."[13] (The *Argo* carried lumber for building the new house on Flaxman Island). A week after tying up, they had not moved. Through cracks in the ice Leffingwell speared 15 small fish for additional food. And each day, he burned 12 feet of two-by-fours in the stove.

Arrival at Barrow and On to Flaxman Island

Late morning August 12, Leffingwell saw a "water sky," indicating open water. At 4:00 p.m. when they sounded their fog horn, the *Karluk* responded, and soon they saw the village of Barrow. Leffingwell wrote: "Very strong N current & at 6pm thought we saw opening in ridge & went thro boiling mass of rubble & slush into a pocket of water then brought up against heavy ridge again. Afraid of breaking propeller & being drifted to N Pole so back & tied up in pocket in ridge near Duck Station. Wind shifting from N to NW & we in very unsafe place."[14]

The crew went ashore for some drinking about the time Storkerson and Gordon, a Scottish whaler from Barrow, came aboard the *Argo* for a visit. Later that evening, they moved the yawl to a safer mooring. The next day, Leffingwell and Storkerson went ashore for visiting and meals and Leffingwell returned by 6:00 p.m. Arey and Gallagher arrived from the east in an oomiak the same evening and Leffingwell stayed up past midnight talking with Arey.

As a result of his conversation with Arey, Leffingwell decided to give up the idea of whaling. He sold the small whaling boat he had purchased from the *Narwhal* to Captain Cottle of the *Karluk* for $250, about what he had paid. The *Thetis* had arrived the same evening, towing the *Jeanette* into Barrow. She was also carrying mail and gear for Leffingwell. Leffingwell hauled two loads of provisions that he had stored the previous year from Brower's place back to the *Argo* and later returned to Barrow for the remaining provisions. Some of the crew had decided to leave the *Argo* at Barrow, which suited Leffingwell just fine.

Early on the morning of August 17 Leffingwell and the *Argo* headed east for Flaxman Island, a distance of 256 miles. Storkerson, Arey, and Gallagher were aboard, the father and son's canoe "lashed to rigging." Travel conditions were good and their route was generally closer to the shoreline. Initially there was thick ice along the reefs, but Smith Bay was clear, and the ice was several miles offshore near Cape Halkett.

By Saturday, August 21, winds were blowing 35 miles per hour and strong currents were against them. Ice was still an issue and their anchor was too short, dragging when they tried to set it. They had reached Cross Island, but had to return to Midway Island, where they finally set anchor. After supper and some whiskey, Leffingwell turned in at 1:00 a.m.

The next morning, Leffingwell intentionally grounded the *Argo* to wait for better travel conditions. He was grateful to be settled in a place without much swell, rather than drifting back to Barrow on strong currents amongst grounded ice and thick ice floes. Snow fell off and on throughout the day; and the wind lessened.

They hoisted anchor on August 23 at 8:45 a.m. and set sail. The engine was not cooperating, but they reached Cross Island in fewer than two hours, and Pole Island by 4:00 p.m. By 9:00 p.m. they could see Flaxman Island and reached the house shortly after 10:00 p.m., concluding three months travel. McIntyre, or "Mack," who Leffingwell had hired to help in the field, was there, and Leffingwell "called it a day," turning in at midnight. Leffingwell soon learned his caches had been robbed and that little remained of value at the house. Among the missing items were pemmican and milk, and some rope and steel cables, items he needed and had counted on. He spent the day unloading gear amid snow squalls and settled in. He wrote: "Moved into bight behind spit W of house. Had intended to build at Collinson Pt. but too late in year & too much outfit here."[15]

The following day, Storkerson, Arey and his family, Wixra, and others were all there to help unload the ship. Leffingwell distributed sacks of flour and other items as payment. They unloaded the cargo to the beach, and the following day Leffingwell and Storkerson carried the gear and supplies up onto the bluff, except for the timber. Arey and his family dropped by later that day and ended up spending the night. Now that Leffingwell was back, there were projects to complete before winter conditions arrived.

Back to Barrow

Leffingwell decided to return to Barrow with Storkerson and Arey to retrieve additional provisions stored the previous year. They traveled west in the *Argo* on August 28 through loose ice, and arrived at Barrow late in the afternoon on August 31. After they anchored near the school, they temporarily cached their food and gear on the beach and headed off to see about retrieving the other provisions. The pack ice was not visible from shore but they carefully monitored the drifting cake ice. Leffingwell and others loaded provisions and lumber on September 1. They departed for Flaxman Island the following day and were plagued by fog, ice cakes, winds, and poor anchoring locations. It was slow going because the *Argo* was weighted down by lumber and provisions.

They took soundings as they went along, not only for anchoring but for the ongoing mapping of the coastline that Leffingwell had previously worked on. Near Pole Island, Leffingwell and Storkerson placed a pole vertically to be used in triangulating the coastline for mapping. They continued past Midway Island and Cross Island, and Leffingwell wrote: "Made running sketch map of islands, soundings etc. Ashore & put pole up on end of hook & piled fire wood."[16]

They finally reached Flaxman Island and the house at 1:00 p.m. on September 11; the round trip to Barrow had taken over two weeks. With the help of Gallagher and another man, they immediately unloaded the deckload of lumber. Over the following two days, with help from

Inupiaq families, the rest of the gear was brought to shore and up onto the bank.

Flaxman Island

Leffingwell set about cleaning the cabin, organizing his gear, and sorting through the boxes. One of the instrument boxes had not arrived, but Leffingwell did not identify its contents in his journal. He started recording temperatures again in his journal on September 15. Snow fell all night and temperatures dropped to -4°. Snow drifted up to and around the house. And over the next two days of blowing winds, ice formed in the lagoon, the ponds were nearly frozen, and ice formed in the buckets inside the cabin.

Leffingwell worked on his plans for the house, including a concrete chimney. He started modifying the cabin, adding a door, boarding up another, and using sod to help insulate where needed. With the help

The yawl *Argo* on the Arctic Ocean. *USGS Photographic Library, Leffingwell Collection*

of McIntyre, he started constructing a new home that eventually connected to the old. He had forgotten roof beams and he was short on two-by-fours, so instead of spacing them at two feet, spaced them three to four feet apart. By September 22, heavy snow was falling, but in spite of the weather they managed to get the rafters up and one-half of the roof completed. By the next day, they had finished the roof amidst a 25-mile-per-hour wind and more snow. Their final tasks were to paper and sod the walls and roof.

In 1909 Leffingwell planned to build another dwelling next to the cabin that had been built from the remains of the *Duchess of Bedford*. In fact, by the time he left Flaxman Island in 1914, he had built a significant and well-designed base camp. Beginning in 1908, he had carefully considered the house design. He determined that the prevailing winds blew from the east and west, which meant that the north and south sides of the house were free of snow and best for windows. He contemplated placing skylight windows in the roof for better light. As he quickly learned, moisture inside any dwelling was a major consideration. He observed where moisture problems were greatest and modified the structure inside and out to reduce the movement of air and make the house as airtight as possible. Heating was a challenge. Leffingwell used stoves for heat but, as he learned more than once, the potential for fire was significant given certain weather conditions.

Connected to the house was what he referred to as the "outer room," used for storage of drinking water, food provisions, dog feed, and firewood. Next to the outer room was a shed to house the dogs, cut firewood, and store the sledges. Near the house lay a cellar used to store meat. When they first arrived in 1906, they had built a rack for provisions. The rack was elevated about five to six feet off the ground, allowing the snow to pass under. Convenience and practicality were hallmarks of Leffingwell's planning.

Temperatures varied from -1° to -10° throughout out the last week of September. On September 30, Leffingwell cut the *Argo's* anchors out of the ice and placed them on the ship, which was not going anywhere. The following week's temperatures were cooler; between -9° and -15°. Leffingwell noted that the stove in the new house worked well. The morning temperature of -9° quickly rose to 12°. But along with the colder temperatures arrived the darkness of fall. He lit the lamp at 5:30 in the evening and the lantern at 6:00 in the morning.

Leffingwell finished papering the north side of the house by October 4. One of his indoor projects was making a stove to exchange for a caribou skin. Another was papering the ceiling for additional insulation. Bunks and shelving were constructed. Soon there were two stoves warming up the new house.

On October 10 Leffingwell walked to Shugvichiak's camp at the east end of the island, noting that it had been a long time since he had taken a walk. When he returned, there were 30 Inupiat at the house for trading. Leffingwell wrote: "Native brought drum & with dish pans etc had dance after supper in new house. Bed 10pm. Temp 0° am."[17]

When Leffingwell was outside he worked on any number of projects, including creating an ice front for the shed and a frame for canvas. The ice front provided insulation and a snow block. He laid a tarp over the shed and banked the north side to keep it in place during winds. He also "Made astron [astronomical] pier out of barrell set in ground & filled with sand & water. Melted holes for posts of observatory walls. Moon & clds eve."[18] When he needed a hole to be dug for a marker, Leffingwell simply built a fire to melt the ground ice, and then planted a post. Later he made a wind break for the astronomical pier.

In the evenings Leffingwell worked on cleaning and calibrating instruments, made star lists to aid in celestial observations, and read. Inupiaq families dropped by and Leffingwell often played the phonograph. Sometimes the families brought "ice fish," which Leffingwell described as four to eight inches in length and tasty but without much meat. One early morning he determined that the temperature on the floor was -2°.

Storkerson left for Herschel Island, a 170-mile trip, on November 3 on a 500-pound sledge drawn by three dogs. Leffingwell wrote that he was sorry not to be going, noting that he had spent too much time indoors. But instead, he continued cleaning and adjusting instruments, took observations, slept, and read. He did not appear to be particularly motivated. Winds blew steadily from the west for a change, and the aurora borealis was so bright he could read inside the house.

For much of November and December, Leffingwell settled into a quiet existence with little company. He described his activities as "loafing" and reading; he studied Inupiaq grammar, but made no astronomical observations. He spent a few days outside "banking" the house and roof; using the snow to further insulate the structure.

Arey and his family arrived on December 20, and that night eleven people slept in the house. Leffingwell and Arey spent the next couple of days talking and catching up. They had a "big feed" on December 24. Christmas Day was a celebration that brought 27 Inupiat to the house for a large meal and dance. The temperature was -30°, and winds were blowing at 36 miles per hour. Due to the winds and cold temperatures, and in lieu of a two and one-half mile walk home, 20 guests slept over.

On December 27, temperatures warmed to -12° and, once again, melting snow from the roof was dripping inside. The winds started blowing the following day and by December 29 were blowing about 50 miles an hour. Leffingwell wrote: "Windows bending in with gusts. afraid they might break. cutting snow off of roof."[19] He also wrote that the wind was "hard to walk against" and that visibility was very limited.

By the end of 1909, Leffingwell was comfortably settled in the new house on Flaxman Island. The winds had temporarily eased up, there

Dwelling and sheds on Flaxman Island. *USGS Photographic Library, Leffingwell Collection*

were snow drifts of a few feet outside, and he could see open water not more than a mile offshore.

A New Year—1910

The New Year brought more cold, wind, and snow. Temperatures varied from -6° to -37°, and winds varied from 15 to 50 miles per hour. Visitors came by most days, sometimes for trading. Arey, who had been staying at the house, left on January 5. A fellow by the name of Sandstrom arrived from the nearby whaler *Teddy Bear* on January 13 (Leffingwell's birthday), and departed on January 15 for Barter Island with a sledge and three dogs.

Leffingwell was busy with a number of projects. He was drawing a map of the lagoon, and working on his "accounts," which were probably trading endeavors and perhaps the status of his provisions. And he was making a new sledge. He took observations when possible and continued adjusting his equipment to yield more accurate results. On January 23, Leffingwell's journal entry was matter-of-fact as usual, but news regarding Sandstrom offered insight into the fragility of life for those unprepared and inexperienced in the Arctic environment. He wrote that the wind was blowing about 15 miles and it lessened in the afternoon. Skies were clear and he continued: "Set alarm for 3am but clds so missed 3.2 mag occult...Terigloo says that Sandstrom did not stop at his tent, also that he T [Terigloo] saw his S [Sandstrom] tracks on young ice. If he S [Sandstrom] was out on ice, prob. lost, for ice broke off that night. No camp & only little grub & poor clothes."[20]

Leffingwell saw Halley's Comet on February 1. He spent three days working on the Kugruak River (Canning) map before he turned his attention to other projects. He had consistently noted that the floor of the house was much colder than the rest, so he made a ventilator shaft designed to transfer the cold up off the floor and out

When Captain Joe Bernard of the whaler *Teddy Bear* arrived with Arey's son Gallagher on February 29, they indicated that the sledge and sleeping bag belonging to Sandstrom had been found, but they had not found his body. Leffingwell noted that, when Sandstrom came to Flaxman, he had not made a snow house but slept out in the open on the snow. Bernard presumed him lost. One set of tracks were headed onto the ice with one dog. Arey had found the sledge. "S [Sandstrom] had no tent & only a little food and did not know enough to make a snow house. Also travelled in dark over strange rout[e], and simply lay down in his bag where ever he brought up. We persuaded him to go back to Barter Id with Terigloo, but he cut out on new ice & missed T's camp in dark."[21]

Spring was returning and Leffingwell noted that it was light until 7:00 p.m. He planned to leave for Barrow on April 1 to try his hand at whaling. It was another attempt at trying to offset some of his expenses. Throughout March, Leffingwell spent much of his time reading, taking and developing photographs, and hosted many gatherings and dances at the house. He recorded temperatures of -15° and -11° on March 16 and wrote: "Beautiful, clear, calm, hot, day. Bareheaded & barehanded, sitting in sun & handling gun barrells...House full of natives, babies everywhere. Two big feeds of caribou meat. Feel like a man again. 11 of us in house at night...Roof dripping inside. 1st thaw for long time. Barom low & rising. Light in house 4am-9pm."[22] It was now March 18, 1910, and Leffingwell wrote that his energy had returned. He attributed the increase in energy to the sun or a meat diet. It may have been a combination of both.

Whaling near Barrow

In anticipation of the breakup of the ice, and possible damage to the *Argo*, they worked at clearing ice around the ship. Towards the end of March, they also spread sand around the ship perimeter to help thaw the ice.

Leffingwell and a large group that included Storkerson, McIntyre, Gordon, and some Inupiaq families departed for Barrow on the morning of April 1. During the trip temperatures were about -25° and Leffingwell's sledge was pulled by four dogs hauling 400 pounds of gear. They traveled 29 miles in eight and one-quarter hours and before they stopped to camp, Leffingwell was stiff, sore, and tired. He had been inactive over the winter months and had exercised little.

They hunted along the way, and spent four days in one of the camps before continuing their journey. The group traveled across the ice of Harrison Bay and several, including Leffingwell, had eye problems that he attributed to the smoke generated by the stove. He bandaged his left eye first, then bandaged both for most of a day for recovery. Off and on they wore Inupiaq-styled sunglasses with narrow slits to limit eye exposure to the brightness and glare of sun, snow, and ice. On April 18, winds blew about 15 miles per hour out of the east and north. They rigged up a sail with a mast and boom for the sledge, and soon were off at four and one-half miles per hour. After they camped in an Inupiaq igloo at 5:30 p.m., Leffingwell wrote: "All pretty tired. Sail great help. No sail we couldn't have made 25 miles...6 lbs flour, 5 bacon & 20 lbs meat left. Wind a god send."[23]

The following day Leffingwell was stiff and sore and had worn a hole through his breeches. It was chafing so he rode on the sledge for some distance. They reached Barrow on April 21, three weeks after leaving Flaxman Island. During the month of May they tried to capture a whale, without success. A lack of open water and leaky watercraft contributed to his frustration. His next journal entry was written in early June.

On June 9, Leffingwell wrote: "dogs took me out on 100ft lane of rotten ice & would not 'whoa'. About 30 ft from hard ice dogs all went thro but sled held me up. Cut one dog loose & hitched sled along backward till I got the others dragged out & they backed to good ice. Got in about mdt Tues."[24] It was another potential disaster avoided by luck. He slept most of the day following his return.

When the Inupiat took a whale, dances were held in celebration. A dance at Barrow started on June 11 lasted three days. On June 15, Leffingwell, McIntyre, and Storkerson left Barrow to return to Flaxman without having taken a whale.

During the return journey, they hunted caribou, ducks, geese, and gathered eggs. The hunting provided their food along the way, and any extra food was stored for the winter and traded. When they made camp, they pulled the boat ashore, turned it on edge, and used the sail as a cover. Temperatures were warming to 2° and 4°. And as a result, traveling conditions were at times challenging because of the changes in ice. It was melting quickly and alternated between rotten ice and water. As they crossed to an island on June 17, they sometimes waded in water that was knee deep.

Weather conditions, especially the wind and fog, dictated whether they were able to travel, and if so, the direction and speed of their travel. On one occasion, when fog forced them to land and make camp, Leffingwell and Storkerson baked a duck in hot sand for an hour and enjoyed the results for dinner. Some days they traveled only a mile, on other days, more than 12 miles. The weather varied, but was often clear, bright, and beautiful. On those days it was important for them to pay attention to the glare from snow, ice, and sun. Leffingwell once again experienced snow blindness in his left eye.

The sledge traveled on the ice or overland, while the boat negotiated the icy water. As they traveled east, Leffingwell also observed the ground ice and geology, including granite and gneiss. Leffingwell recognized that the boulders came from far away, but lacked current understanding of their deposition.[25]

Leffingwell determined that their ammunition supply for hunting was low and their food supplies consisted of 18 pounds of flour, 30 pounds of beans, and some tea and coffee. They cached the sledge at Cape Halkett and continued their journey by boat.

Their rate of progress increased when McIntyre rigged a "balloon jib" from the sledge cover. Leffingwell wrote: "Breeze up & we tore along for 2 hours @ 6 miles."[26] When they reached Oliktok, the mosquitoes were waiting. Leffingwell used moss to build a "smudge" of smoke to discourage them. But they were not discouraged for long. The following day the mosquitoes were biting through his fur shirt. He removed his undershirt and placed it over his head for protection.

At Oliktok, they camped near a pingo and set up a tripod and signal for mapping purposes. They also took soundings along the way for navigation and for Leffingwell's mapping of the coastline. When they arrived at Pole Island on July 12, Leffingwell observed that the ice pack was moving to the inside of the Island.

Flaxman Island

On July 13 the party arrived at Flaxman Island about 8:30 a.m. after six hours of travel. When they arrived, Leffingwell wrote: "Sagi [Shugvichiak] & 3 other tents on beach in front. Argo safe thanks to Sagi, leaked badly & he bailed water…for a long time…Things in fine shape thanks again to Sagi. Only 7 hours sleep since Mon morning."[27]

Leffingwell and others worked the nets, taking in fish, and he worked on his accounts of provisions and trading. He planned to return to Barrow, load more supplies, take soundings and establish signals and beacons for mapping of the coastline. He started the mapping on

Flaxman Island and, with Storkerson's help, established an azimuth mark on the mainland. On Flaxman, Leffingwell cast several concrete piers to assist in astronomy and mapping. He was working long hours and made good progress. Leffingwell was engaged in his work once again. On Wednesday, July 27, he wrote: "Put up Δ [triangulation] mark on hummock & surveyed, while boys tended nets & shot a few ducks. Mapped edge of it & 1st opening & mainld 2 miles S.E. Finished work noon & after lunch nap 1:30-4:30. Started back 5:00...turned in 10pm. worked 30 hours straight. 3 hours sleep in pm (rate of traverse, pacing = 1 miles in 1 1/4 hours). Couldn't sleep till past mdt."[28]

To Barrow on the *Argo*

On Monday August 1, Leffingwell, McIntyre, and four dog pups left Flaxman at 11:00 a.m. for Barrow aboard the *Argo*. They erected signal poles for his mapping of the coastline along the way and Leffingwell mapped some of the islands en route, but the purpose of the trip was to retrieve more provisions.

They also set nets for fishing when and where they could. The nets yielded more fish than they ate so they used racks to dry and salt the fish. Some of the fish were placed in sacks and buried for future use as dog feed. They also cached two nets for future use.

Leffingwell and McIntyre arrived at Barrow on August 10, loaded provisions, and departed on August 13 for Flaxman Island. They arrived at 4:00 a.m. on August 16. Provisions and more lumber were transferred from the ship to shore. Leffingwell and Arey then made a trip to Arey's cabin at Collinson Point and unloaded gear and provisions belonging to Arey.

Leffingwell was back on Flaxman by August 24. The whaling ship *Herman* had dropped off mail and a box of oranges from home. He spent the next couple of days getting ready for another trip to map the coastline. When the *Karluk* arrived on August 27, Leffingwell acquired some pork and sent out some letters with the ship.

Leffingwell had left Seattle in May 1909 and reached Flaxman Island on August 23, 1909. A year later, his accomplishments were scattered and fell short of his expectations. He had spent much of his time hauling provisions to Flaxman from Barrow, a distance of about 256 miles. He had worked on further construction and improvement of the

Flaxman base camp. He spent three and one-half months trading and fishing, and making an unsuccessful attempt at whaling. While traveling between Flaxman Island and Barrow, he established additional triangulation stations and surveyed. He had constructed the astronomical pier near the Flaxman house, had taken observations when he could, completed some mapping, and worked on the ongoing cleaning and calibration of his instruments. It was time to prepare for the next chapter, and finish the mapping of the coastline and interior.

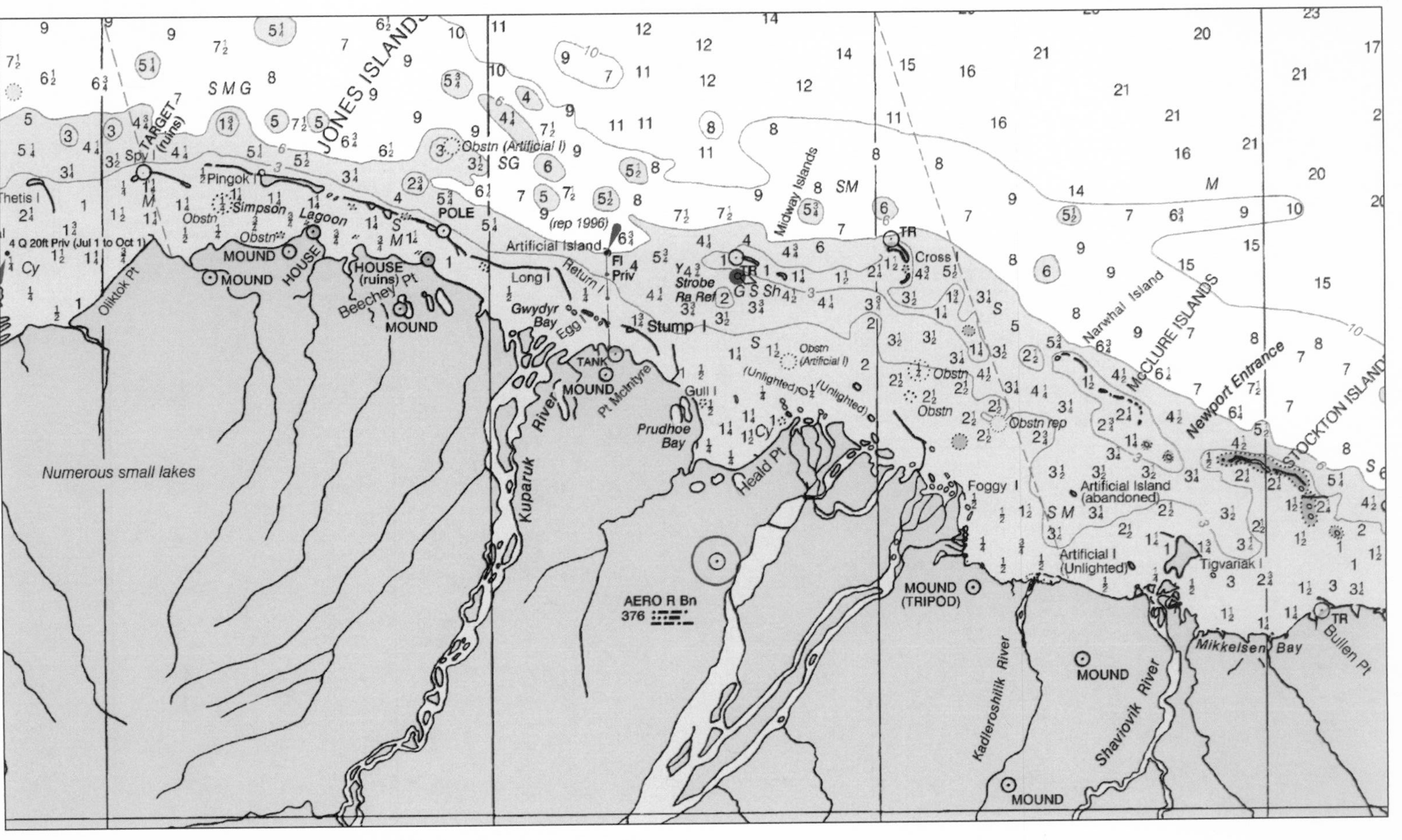

Northern Alaska coastline; Oliktok Point to Mikkelsen Bay. *NOAA nautical chart 16004.*

CHAPTER 9

Mapping Coastline and Mountains, 1910–1911

Leffingwell's continued mapping along Alaska's Arctic coastline required even more resiliency than his work in the mountains. There was seemingly endless checking and refining of results for accuracy. Triangulation stations Leffingwell established were often destroyed by the elements of wind, water, and erosion.

On Wednesday, August 31, Leffingwell and McIntyre again embarked on the *Argo* to begin mapping the coastline from Flaxman to Barrow. It would be one of many such trips. With them were two dogs and four pups. They stopped on the mainland long enough to put up a pole as a beacon for triangulation, the basis for his mapping. By using the triangulation stations to measure angles from other locations, Leffingwell achieved great accuracy in his map details. For two weeks he and McIntyre mapped the islands of the coast. As they continued west, they placed each day a pole that varied in height from 20 to 35 feet. Each pole had unique cross boards for identification, and Leffingwell drew and identified their location in the back of his journal.

Leffingwell's published report detailed and described over 50 triangulation stations starting from the base at Flaxman Island. He determined the location of that base from astronomical observations. He indicated station locations with physical descriptions, latitudes and longitudes, and whether a subsurface marker was placed, in case the surface marker was destroyed by weather or waves. His general description and mapping of the coastline was completed using a plane table and field sketches.

On September 9 they were at Midway Islands north of Prudhoe Bay. Leffingwell finished work and they were off on the *Argo* to two additional islands further west before stopping to camp. It snowed on September 13, so Leffingwell decided to quit mapping the islands. He and McIntyre turned their attention to the mainland and the mouth of the

Kuparuk River, where they hoped to catch some fish to feed the dogs and cache some flour. The following day they decided to place a pole on a pingo they had seen upstream. For the first time that month the pools had frozen over. They reached their destination after some poling of the boat, and some sailing. It had been a long day; Leffingwell's pedometer recorded 37,000 steps taken, and he had lost his mittens. Upon arrival, he drew a map and description of the cache location, and continued mapping some of the surrounding area. On September 16, while they ate lunch on the west side of Prudhoe Bay, it snowed again. Another couple of miles were mapped before they placed a signal pole. The wind picked up over the next few days and on September 19 Leffingwell noted increasing slush in the water. By September 20 it was obvious that an Arctic fall on the North Slope of Alaska had passed quickly. They had been mapping for three weeks and decided to return to Flaxman Island. Leffingwell observed a flock of loons heading west. Winter was on its way.

Flaxman Island

Their timing was good as a gale blew in on September 21. The following day, Leffingwell piled wood from the shoreline on the bank and filled the water barrel. He wrote letters for the Inupiat to take south to Fort Yukon. He then traded for nine caribou skins, and removed the double-glassed windows from the house to dry them in the sun.

With little time left before winter, Leffingwell and McIntyre began building the 12-by-14-foot storehouse. The gaps between the boards were filled with pitch. The south side was then painted with a hot seal oil. By September 25 the storehouse was nearly finished except for a door and some work on the roof. Leffingwell recorded temperatures between 0° and 3° and observed red-throated loons heading west. The final days of September and early October brought snow flurries and some accumulation of snow. Using the sledge they moved food into the storehouse and finished other outside tasks, including hauling water. On October 2, he wrote: "Ponds frozen nearly over."

As fall quickly became winter, Leffingwell observed the "old ice" visible offshore. He paid careful attention to the fluctuations of the barometer and recorded the temperature twice each day. During October and the first half of November temperatures were usually about -10° but fluctuated between -3° and -24°. On October 25 the sun set at 3:45 p.m.

Leffingwell and Shugvichiak, whose camp was on the east end of Flaxman Island, were good friends and visited regularly. Shugvichiak's family had been living on the island when Leffingwell and Mikkelsen arrived in 1906. His boys (one was Kissik, the other's name is not known) were learning English from Leffingwell, and Leffingwell was learning the Inupiaq language from Shugvichiak and the boys. Shugvichiak's wife Toklumena frequently made or repaired Leffingwell's clothing in exchange for food.

On October 31, McIntyre and Chitsak left Flaxman Island on dog sledges for Herschel Island. They carried letters written by Leffingwell. Meanwhile Leffingwell delved into projects. He laid a new floor in his room, made a new darkroom, a desk, a new bunk, shelves, work bench, and tore out his old darkroom, which added three feet to another room. He even painted the walls of the new space. Leffingwell also worked on new dog harnesses and caught up on

Shugvichiak (Sagi) and children. *USGS Photographic Library, Leffingwell Collection*

correspondence. He had outside projects, too. He transferred lumber from the shed and old house to the outside rack. The boats were moved from the shore to the top of the bank. Leffingwell noted there was ice near shore, and slush in a calm ocean.

Leffingwell was generally good natured but had a temper. He was also very fond and protective of his dogs. Leffingwell neither owned the land, nor could he control subsistence hunting, but he was adamant about protecting the dogs, which he refused to tie up. At feeding time on November 8, he noticed that one of the dogs was missing: "& calling, heard him howl. Went at run & found his foot in wolf trap. No 3 set by Okalishuk for fox 1/2 mile from house. Bait lying around. No bones broken, balls of feet & toes frozen. Don't think he will lose any toes. Very angry. Took trap & told boy to remove all traps from island or I should use an axe on them."[1]

About once a week Leffingwell took the pups out for training to pull a sledge; he was pleased with their initial performance. He modified one of the sleds by trimming four feet off an eleven-foot sled and was pleased with the resulting weight: 20 pounds. When Leffingwell tried the new sledge with the pups, he found them very stubborn. They did well until they wanted to come home, at which point they would not stop, no matter his commands.

The next day the anemometer indicated winds of 62 miles per hour, but Leffingwell felt that they were likely higher. The lumber was blown off of the provision rack, and the winds blew through cracks into the house, so he cemented up the vulnerable places as best he could.

Toward the end of November he switched from his sleeping bag to blankets, but changed his mind after one night. Conditions were quite different from those in Franz Josef Land, where he slept with minimal blankets and stayed warm.

By December 3, McIntyre had still not returned from Herschel Island and Leffingwell noted that he had been away for five weeks. The weather deteriorated and his concern for McIntyre grew. Temperatures were consistently in the -30s and high winds were constant. Leffingwell theorized that McIntyre had been delayed by water, or that he was at Arey's, or that he was on his way, arriving in moonlight. The weather was poor for eight days and in spite of the weather, Shugvichiak and his son had come for a visit, thinking that Leffingwell might be lonely.

Leffingwell appreciated the company. Leffingwell finally received word from Arey that McIntyre was with him. As much as he cared and worried, Leffingwell's sense of relief turned to anger and resentment: "Damn Mack, I want to get to work on maps etc & I can't do all the house work & have time for...work also. Spent whole eve till mdt. getting map gear ready etc."[2]

McIntyre arrived about 3:00 p.m. on December 19, after 11 days at Arey's, and six days returning to Flaxman. Travel had been difficult as the sledge kept overturning. He got lost in the fog, and came across a body encased in ice. Leffingwell wrote: "Prob. Sandstrom, lost from *Teddy Bear* last year. Clothed that way. Frozen in to cake of ice, legs only exposed... Mack says he nearly lost his nerve when he got lost in fog. Thought of Sandstrom's body."[3]

With McIntyre home safe and sound, Leffingwell worked for the next five days on maps of the coastline and offshore islands. He added soundings to the maps he had obtained on previous trips and worked on a plan for triangulation of additional coastline. He also spent time in the observatory trying to get occultations.

Christmas passed unobserved, except for Leffingwell's mention that he "Got a feed for natives." Only one person showed up on December 25, but Shugvichiak and Terigloo's families were over all day on December 26. The women dropped by on other days with newly sewn clothing for Leffingwell and McIntyre: kamiks, breeches, and atteges for winter. On December 29, 1910, Leffingwell described seeing the mountains to the south for the first time in many days. Early January brought temperatures in the -40s, and there was frost on the door, which was frozen shut.

Shugvichiak arrived on January 17, feeling sick, with a pain on his right side. Leffingwell thought his friend looked a bit jaundiced and that perhaps it was his liver. He gave Shugvichiak some medicine but did not record what it was. Two days later Leffingwell visited Shugvichiak and found that the area was swollen and was sore when touched. He was still sick when Leffingwell visited him again eight days later.

Leffingwell had another close call with fire in the house when soot in the chimney pipe leaked and sparked. A hole burned through a nearby board before the fire was extinguished; he then managed to kick the chimney off the roof. He had been at the Flaxman house since

September 20 and over four months had passed. It was early February, and after completing projects that included packing books to be shipped home during the summer, cleaning and calibrating instruments, and making observations, he prepared for another trip up the Canning River.

Return to the Canning River

Leffingwell planned to establish more stations for mapping. Along with McIntyre he left Flaxman Island on February 11, and stopped near the coast to establish a marker on a pingo before heading up the Canning. They reached Ignek Creek on February 18, Shublik on February 20, and Arey's old house south of the Ikiakpaurak Valley on February 21. They spent 10 nights in the area while Leffingwell surveyed and took angles, and returned to a station located on a nearby hilltop he had established previously.

After 300-plus games of solitaire and no luck at fishing, they began the return trip to Flaxman Island on March 3, and established camp 8 at Shublik for three nights. Between snow and heavy winds Leffingwell spent next two days unsuccessfully looking for a suitable triangulation station location.

Tent after a heavy gale, Arctic Ocean. *USGS Photographic Library, Leffingwell Collection*

Leffingwell and McIntyre spent the following five nights at Ignek Creek. Along the route to the creek, Leffingwell found an outcrop of fossils and collected about 100 pounds of samples in an hour. He hauled the samples to camp on the sledge, cracked them open, and described them as "K. gray Qt.," Cretaceous gray quartzite.

The weather took a turn for the worse; gales exceeded 50 miles per hour with heavy drifting snow. Temperatures were in the -30s and-40s. Leffingwell wrote: "Blowing a little harder than yest[erday] and colder. Burning green willow and tent cold all a.m. Top frosted inside...Mack rustled a load of dry wood after noon and got tent warmed up. Coldest blow in my experience. I didn't go out at all. Only time I have seen frost inside of tent with fire going."[4]

Leffingwell and McIntyre reached Flaxman at 4:45 p.m. on March 13, having been upriver for a month. The house was extremely cold and the storage shed held a three foot snow drift just inside the door.

Two weeks later, Leffingwell headed out again. He prepared by making a list of the type and quantity of food he and McIntyre had used, and another of the feed used for the dogs on the previous trip up the Canning River. In addition to the practical lists, he spent part of this time "planning castles in Pasadena." (It was a reference to property his father owned in Pasadena, California, where Leffingwell hoped to build.) There were visitors, trading, and the usual cleaning and calibrating of instruments. One of the visitors was Arey's son, Gallagher, who had been off hunting and had recently returned to his family, only to find that they were suffering from scurvy.

Another Trip West

McIntyre convinced Leffingwell that he should try a pyramid style tent on this next trip. Leffingwell cut the canvas and McIntyre did the sewing; it took about a week to complete. They set up the tent outside the house and were favorably impressed—they decided to take it on the upcoming trip. McIntyre visited Shugvichiak on March 24 and reported that he was still sick. Two days later Shugvichiak visited Leffingwell, but by March 31 Toklumena reported him very sick. Leffingwell speculated it was a kidney condition, but when he visited Shugvichiak, he knew that it was the liver because Shugvichiak was jaundiced. Leffingwell probably knew he could not help Shugvichiak, and likely understood the inevitable outcome.

On April 2, Leffingwell and McIntyre set off at 6:00 a.m. and headed west with dogs and a sail rigged to the main sledge to continue surveying the coastline and islands. They also carried a small sledge for hauling the instruments. He had arranged for some of the local Inupiat to hunt for them, so they could focus on the surveying and mapping. The Inupiat had already left Flaxman to start hunting.

Leffingwell's first camp was on Challenge Island. Traveling west, they established more beacons and signals for triangulation, taking angles and soundings as they went. They numbered or named most of the islands west of Flaxman Island; most of the names are in use today. Some of the place names honored Mikkelsen, Storkerson, McIntyre, Sweeney, Stefansson, and Arey. Others honored revenue cutters and whaling ships, including the *Thetis, Duchess, Karluk, Jeannette, Narwhal, Belvedere, Alaska,* and the *Mary Sachs*, and their captains, including Cottle, Leavitt, Bertoncini, and Bodfish.

Leffingwell and McIntyre met up with the hunting party on April 12 near the Kuparuk River and from then on they traveled together. The trip lasted a month and extended as far west as Oliktok Point before the return to Flaxman Island. Generally they camped only one night at each location unless forced by weather conditions to stay longer. After several days hunkered down in tough weather, Leffingwell decided that he did not much like the new tent, especially the pole in the center, which restricted their movement.

Flaxman House and Shugvichiak

On May 3, they were up about 11:00 a.m. and started for home about 2:30 p.m. The winds blew from the west about 15 miles per hour and they hoisted the sails on the sledges. Making good time they arrived back at the house at 8:15 p.m. Everything was fine except the winds had again blown some of the lumber off the rack.

They had been gone a month, and the morning after their return Toklumena visited to report that Shugvichiak's condition had worsened. Leffingwell focused on instruments, worked on his accounts, and did some cleaning, then started preparing for the next trip, this time to the east. They planned to depart on May 10. Toklumena returned again on

May 9, and indicated that Shugvichiak was bedridden and unable to eat. The next day one of Shugvichiak's sons dropped by the house with the same message. Leffingwell wrote: "Off 5:45. Stopped at Sagi's a half hour. Old man nearly gone...Only thing I can think of is abscess of liver & perhaps Brights [kidney disease]...Felt ashamed of not having been down before, but had no idea that he was any worse. Sunday he was trading with bunch that came...Don't think he will live more than a week-4 days. Glad not to be around for I can't do anything, he can't eat, even Malted Milk."[5]

Leffingwell had been so focused on trip preparations he had failed to recognize the visits by Shugvichiak's family as a plea for his help, or how sick Shugvichiak was. After a brief visit, he continued his journey toward the Sadlerochit Mountains.

The Sadlerochit Mountains and Schrader and Peters Lakes

After traveling 10 miles eastward on foot, Leffingwell and McIntyre established their first camp about 10:30 p.m. on an island in the Canning River delta. The following day they continued on to Konganevik Point, where they were forced to stay seven nights because of high winds, drifting snow, and poor visibility. On two days temperatures were recorded at 0°. Other Inupiaq families arrived and Leffingwell counted nine tents in the vicinity. He spent the hours reading and taking walks. On May 17 the tent was almost buried in snow drifts and it was a challenge to get in and out. By May 18 he'd had enough:

> Up about noon, blowing 40+ miles as usual but snow only slight drifting. Thot [sic] we'd get away, but 6pm, blowing 50 miles & heavy drift, snowing?...will have to prop up inside tonight. Dripping most of day. Sticks all askew from pounding & top slow sinking. Have to crawl on belly to get in & out & snow wet. Floor wet. Feet wet & cold. Wood wet. stove wet. The rottenest camp I remember. At Ned's old house in Feb. could walk around outside. Here no go. Tried it today but too hard to push against wind. Chunks of snow flying in face.[6]

Leffingwell walked east to Collinson Point where he, McIntyre, Aigukuk, and Karoiga met up with Arey and others. They followed Marsh Creek toward the Sadlerochit Mountains, traveling 21 miles in 11 1/2 hours. It was now May 23 and Leffingwell's attitude was much improved: "One mile from winter to summer. Sun shining in tent door at mdt. as we ate supper. Warm & fine."[7] The temperatures, which went from -2° that day to 9.5° the next, certainly reflected his thoughts. Leffingwell was up at 10:00 a.m. and started sorting gear for the packs that he, Aigukuk, Karoiga, and the dogs would carry. The following day he mostly stayed in the tent. He made a carrying case for a hand level and a storage bag, and played solitaire. Temperatures remained about 8° over the next several days, but weather was constantly changing from light rain to cloudy to clear and sunny.

Leffingwell, Aigukuk, and Karoiga were about to explore and map the Sadlerochit Mountains, and the lakes to the south.

They set out about noon on May 26 for the eastern Sadlerochit Mountains and Sunset Pass carrying 50-pound backpacks containing flour, pemmican, milk, pork, sugar, and tea. The food would not sustain them for the entire time they would be away, so Leffingwell and the others planned to live off of what they hunted. The dogs carried saddle packs that weighed about 30 pounds.

Sadlerochit Mountains looking south toward Sunset Pass, 1992. *Janet Collins Photographic Collection*

The Sadlerochit Mountains are "a linear range that extends east-west for about 60 miles," divided by a north-south "broad open valley that is called a wind gap and is not a high rugged pass."[8] Itkilyariak Creek originates and flows through that valley. It is quite small and easy to step across. Sunset Pass is located at the southern end of the valley.

They arrived near the Sadlerochit River about 9:00 p.m., but there were dense willows to bushwhack through, so they decided to camp above the river. They spent four nights at the camp. During the days, Aigukuk and Karoiga hunted successfully for ptarmigan and Leffingwell hunted for fossils in what he referred to as the Ignek Shale. Leffingwell found an abundance of fossils and wrote that they "mostly fell to pieces," but he did find some worth keeping, and they were eventually shipped to the Smithsonian in Washington, DC.

The party reached Neruokpuk Lakes south of the Sadlerochit Mountains on May 31. Leffingwell transliterated the name "Neruokpuk" from the Inupiaq word meaning "Big Lake." He decided to name them Lake Peters and Lake Schrader for a topographer and a geologist with the U.S. Geological Survey.[9] (In his journals, Leffingwell referenced the

Aigukuk (left) and Okalishuk on Arey Island. *USGS Photographic Library, Leffingwell Collection*

northern lake as Peters and the southern lake as Schrader but mapped them in the opposite. As a result, today the northern lake is known as Schrader and the southern as Peters.)

They camped on the east side at the mouth of a creek on the northern lake and settled in. Leffingwell wrote: "Fog banks rolling up valleys...Pack hurt head today, but legs ok. From natives, I imagined our route today would be easy, but a "good road" means "passible" with them. Fresh bear tracks on snow bank near camp."[10] Leffingwell's pack was primitive by today's standards and included a head strap to help distribute the load. The shoulder straps were not padded and cut into his shoulders.

They ate ptarmigan for dinner and the next day tried fishing. One of the Inupiat caught five while Leffingwell, using an iron hook, did not catch any. The fish were few and far between, and Leffingwell described them as "poor eating." He was disappointed because he had counted on the fish to supplement their food supply.

After two nights at the northern lake, they moved to the southern or upper lake, now known as Lake Peters, for an additional four nights. Leffingwell started mapping the following morning. With his hob nail boots and 25 pounds of equipment, he hiked to the top of a nearby mountain and after mapping, built a cairn as a signal.

The beauty of Leffingwell's work is that many of his descriptions of geography and geology could be used today as a field guide to understanding the area. He calculated the southern lake (Peters) to be four miles long and one mile wide and located within a U shaped trough. It drains to the northern lake via a narrow channel because an alluvial fan constricts the flow. From the northern lake (Schrader), the outlet was dry at times when other areas were still flooding, so Leffingwell concluded that the water flowed below ground and resurfaced in a side stream toward the main river.

In his report, Leffingwell described the Sadlerochit River as about 80 miles in length, with two forks: a main river or west branch, and the other draining from Peters and Schrader Lakes. He observed that the two forks joined and the river flowed along the south side of the Sadlerochit Mountains before curving north around the eastern end of the mountains. The delta area was about one mile wide and consistent with other deltas composed of "mud flats and silt dunes."

As he observed the surrounding area, he determined that it had once been under one of the largest glaciers in the region. Leffingwell calculated its size at 400 square miles. He concluded that in some areas the ice had reached 1,000 to 2,000 feet in thickness and that the "glaciers did not reach the floors of the main branch" of the river.

By late evening on June 6, Leffingwell had finished mapping around the lakes. June's 24-hour daylight gave him the opportunity to work when the weather was good and hunker down when it was not. Leffingwell often worked in three to four hour blocks of time and then would take breaks for several hours. The daylight meant that some of the working days were especially long. Now that the mapping near the lakes was completed, and running low on food and ammunition, Leffingwell made the decision to leave. The weather was beautiful, he took photographs of Mt. Chamberlin from camp, and regretted spending any time sleeping when he could have been working.

They set off the evening of June 7 and reached the Sadlerochit River about 1:00 a.m. The river was muddy and running high due to spring runoff. Leffingwell struggled with the current, and decided to head directly to the nearest riverbank instead of following the river bar. He went in up to his waist, then tripped as he neared the bank, and ended up wet to his armpits. The Inupiat were more fortunate and managed to stay dry just above the knees. Leffingwell wrote that the dogs had been carried downstream some distance before reaching shore. Leffingwell stripped off his lower clothing and wrung them out. They camped shortly thereafter and built a fire to dry out. Soon the weather turned to snow and the wind increased to 25 miles per hour. Leffingwell wrote: "Nose & head cold in tent. Boys cold in blankets... Now is time when living on game is no fun. Would give a lot for stove & pemmican."[11]

The Inupiat headed out to hunt after midnight. Leffingwell decided the trio should return to the cache on the north side of the Sadlerochit Mountains without finishing mapping within the mountains. Their food supply was dwindling. Without telling Leffingwell, and because he was not sure he could do it, one of the young men returned to the cache for additional food, including flour, sugar, and pemmican. He traveled 30 miles in 15 hours, a remarkable feat given the terrain. The other Inupiat retrieved fossils while Leffingwell traveled upriver to look for a geology hammer he had left behind.

The extra food from the cache enabled Leffingwell to continue his work. He located more fossils from the Shublik Limestone similar to those collected in 1908. His sleep was unsettled and he dreamt of bears and earthquakes.

Leffingwell continued mapping between the rain, sun, and cold winds, and took some photos of the Inupiat and dogs. They traveled six miles back through Sunset Pass to the north side of the Sadlerochit Mountains along Itkilyariak Creek where he continued mapping. Between breaks for weather and food, he worked until 1:35 a.m. the following morning. Leffingwell concluded that the Itkilyariak Creek, which flows through the break in the Sadlerochit Mountains had not been glaciated. The wide open area could not have been created by the creek. Leffingwell thought it probable that the Sadlerochit River flowed through there previously.

By June 13 their food supplies were low again. They had been in the field for close to three weeks and Leffingwell wrote of distant thunder and rain. As they headed north through Sunset Pass he continued mapping, eventually reaching the headwaters of Marsh Creek.

At Marsh Creek, Aigukuk headed to the Katakturuk River, a distance of about 18 miles, looking for caribou, returning 24 hours later without success. Leffingwell headed south along Marsh Creek and into the mountains investigating the geology and found more fossils. He wrote that the geology was complicated, some new rock types were present, and he could not figure it out.[12] They spent five nights at the camp while Leffingwell collected more fossils and mapped the area as best he could.

An Inupiaq family camped nearby and shared news from Flaxman Island. Not unexpected, shortly after Leffingwell had departed for the Sadlerochit Mountains, Shugvichiak had passed away. Leffingwell had lost a generous and supportive friend whom he would greatly miss during his remaining time on the North Slope.

There was other news; two white men had been asking questions about a couple of miners who were lost up the Canning River. Leffingwell and the Inupiaq family shared food, including what Leffingwell referred to as "wild potatoes," flapjacks, and ptarmigan. During their final day at the head of Marsh Creek, Leffingwell and the Inupiat climbed to a nearby station he had established. He took angles, then the three built an eight-foot cairn with an eight-foot base before returning to camp.

Collinson Point and Flaxman House

Leffingwell, Aigukuk, and Karoiga reached Collinson Point late on June 19 and joined Inupiat camped there. They enjoyed a large meal of seal, ducks and tea. The next day they traveled 12 miles to Arey's camp near Konganevik Point and established camp. Leffingwell visited with Arey for hours, but left for his next camp—three and a half miles west at the mouth of the Canning River—the following afternoon. He spent the next few days mapping the coastline and then moved camp to the western side of the delta where he put up a signal atop a pingo. Leffingwell continued his mapping west to his stopping point from the year before and finally reached Brownlow Point that evening at 10:00 p.m.

Leffingwell crossed to Flaxman Island on the evening of June 26 and arrived at the house about midnight. McIntyre was there and during Leffingwell's absence had been caulking and painting the *Argo*. Leffingwell spent the next two days working indoors. "Up 8:00 & worked all day till 9pm & 11pm. letters & map for Brooks," a reference to Alfred Brooks of the U.S. Geological Survey, for whom the Brooks Range is named.[13] By the following day, he wrote: "Finished nearly everything." With Aigukuk and Karoiga's help, Leffingwell had accomplished a significant amount of mapping and description of geography and geology.

To Barrow and Return

After being home for just three days, on the afternoon of June 29, Leffingwell, McIntyre, Aigukuk, and Karoiga started off with six dogs, a dog sledge, and a 22-foot boat toward Barrow. The plan was to map more of the coastline and outer reefs en route to Barrow, where they would pick up provisions that were arriving on the schooner *Transit*.

Upon departing from Flaxman Island the party had situated the boat on the sledge for traveling over the ice. The sledge broke shortly thereafter, so McIntyre returned to Flaxman with the sledge and three dogs. Leffingwell and the Inupiat alternated between lining the boat in shallow water and hauling it over the ice. They traveled a total distance of three miles and finally established camp. Leffingwell then mapped the area from 8:00 p.m. to midnight before returning to camp.

The next day Aigukuk returned to the Flaxman house for tea while Leffingwell and Karoiga set out to continue mapping. Leffingwell later wrote: "Finished to Δ [triangulation] West post at 9pm, only 3-miles direct, but much meandering around bays etc. Map not very accurate but has checks every mile or so. Karoiga made a map of same stretch as I on alone no calc. [calculations] using a ruler to point with & guessing distances. Very good & more detail than I had."[14]

Leffingwell, Aigukuk, and Karoiga worked long and hard hours; Leffingwell mapped as the others hunted, fished, and assisted him. Sometimes they worked over 30 hours at a stretch, but then slept over 20 hours. The 24-hour daylight of an Arctic summer allowed them to work as long as they wished. And there were also days when much was accomplished in spite of heavy fog and wind. Leffingwell established triangulation stations with posts and rocks, took angles, and made field sketches as they traveled.

Similar to the previous coastal trips, the party generally camped in the same location for one or two nights before moving on. If the weather was bad or mapping complex, they spent three or four nights. At times there was enough open water to sail, but at other times they poled the boat through shallow water or hauled it over ice. At the Kadleroshilik River they met three men who helped haul the boat over ice, and they encountered thick mosquitoes for the first time that summer. While Leffingwell and the Inupiat endured the pests as best they could, the dogs responded by cleverly digging deep holes in the sand.

A typical working day involved traveling from island to island mapping. When it rained they stopped and used the sail as a shelter. It was exhausting work. They reached Cottle Island, which Leffingwell had named for a whaling captain, on July 18 and settled in for a rare four-night stay. It may have been the constant companionship, the driven work pace, or Leffingwell's rigid structure, but early one morning Leffingwell lost his temper with Aigukuk. Leffingwell was lying in the tent and an exhausted Aigukuk was half asleep tending the fire. Leffingwell noticed that the teapot was boiling and asked him to pull it before it dried and ruined the pot. Aigukuk refused, indicating that it would not be a problem. Leffingwell told him to do it or else be placed ashore to wait for an Inupiaq boat passing by. Leffingwell wrote:

> He still refused, so I took it off myself...Boys were not sure Barrow natives would stop at Oliktok, if fair wind & Ai might get left. Had his flour & tea all ready for him & said he could use net cached at Oliktok, 20 miles, & would put him on main, if he was still in same mind. Held his nerve & said it was up to me. Dropped mine, & said he needn't go if he understood that he must do as he was told. He said he was angry when I got angry last night. No good that way. Guess he is right. He was half asleep sitting up & I got tired of waiting for him to come to life & let out on him. Too good to loose [sic] unless he gets uppish again. K up like a shot, when I called him this am. Boys off 8pm to main [mainland] with net...Put up stove & made Δ ion [triangulation] plans for around here.[15]

Leffingwell had been raised in a structured environment at home, school, and the military, and tried to impose that structure on his Inupiaq assistants. He realized that there were times when his behavior was inappropriate. Aigukuk and Karoiga had been a tremendous help to him hunting, cooking, and mapping, and he knew he could not accomplish much without their help.

They moved past the incident and continued the mapping, arriving at Oliktok Point on July 23 after having traveled 20 miles in 16 hours. The party camped for three nights, working when they could between the wind and rain, but Leffingwell was aggravated by the weather. At one point, he wrote that he had given up mapping because of the fog and rain. However, Aigukuk and Karoiga had been especially busy fishing and drying fish ashore on racks. Thanks to their efforts, Leffingwell was able to cache several hundred pounds of fish in barrels for future use.

At the next camp west of Oliktok it was stifling hot, and the mosquitoes were so bad that Leffingwell sewed up the tent door. He wrote: "Mapped only 5 miles & have been chewed by mosqs for 23 hours. Too long, got out of temper while sighting toward end."[16]

About 5:00 p.m. on July 30, Leffingwell sighted McIntyre and the *Argo* heading west toward them. He decided to return to Oliktok, and started mapping his way east until the fog rolled in. Aigukuk and Karoiga helped establish signals and subsurface marks, in case the above-ground signal blew down. They reached the *Argo* about 1:00 a.m. They camped at Oliktok Point for five nights, mostly because of heavy winds. On the *Argo*, Leffingwell set a second anchor and he and McIntyre took turns at watch, while Aigukuk and Karoiga camped on shore near other Inupiaq camps. Aigukuk and Karoiga then headed east with other Inupiat to Flaxman, while Leffingwell stayed aboard the *Argo* with McIntyre.

The winds had dropped to 25 miles per hour when they departed Oliktok Point on August 5. Under a full sail and jib, they made good time and arrived at Barrow the following day. The *Belvedere* and *Transit* were there, and the *Transit* was carrying a load of provisions for Leffingwell. They took on 25 tons of cargo and cached the remainder on shore for the next trip. They were underway the morning of August 8 and by late afternoon were cruising in squalls at over 8 knots. On August 10 they anchored close to Cottle Island and took turns at watch. They endured high winds and heavy swells; twice water came over the side. Concerned about their safety and passage, Leffingwell did not sleep well that night.

When they finally anchored back at Flaxman Island near the house at 3:00 p.m. on August 11, the winds had died down. The cargo was unloaded onto the beach the following day, and two days later, McIntyre and Arey departed for Barrow in the *Argo* to retrieve the additional provisions. After they left, Leffingwell covered the provisions on the beach with tarps and moved some gear to the storehouse. He spent the following week sleeping, reading, writing letters, taking walks, and visiting with local folks and others who passed by on their way along the coast.

Mapping Flaxman Island

After taking a week-long break Leffingwell began detailed mapping of Flaxman Island, starting at the west end of the island. He and Aigukuk attempted to map the coast to the mouth of the Canning River, but fog forced them back to the island. He returned to the east end of the island the following day, but full mapping of Flaxman Island would not be completed for a couple of years. He expected the *Argo* to return by August 26, and grew impatient when it still had not arrived on August 28.

The following day Leffingwell discovered he was off by a day in his journal; it was actually August 30, not August 29. The *Argo* finally arrived about 1:00 p.m. A heavy surf at Barrow had prohibited them from loading from the shore and the cache at the Barrow beach had almost washed away. Head winds also slowed their return.

Winds were still blowing strong the following day at Flaxman Island, so with the help of a group of Inupiat from the mainland, the cache was moved up the bank. By 5:00 p.m. they had unloaded half of the *Argo's* cargo. They finished the following morning before noon and Leffingwell then moved the *Argo* a short distance to the east from the house and set anchor. He later wrote: "Paid natives off = one sack flour for each man & 1/2 for women & boys. This is very high, but new trader at Col. Pt. [Collinson Point] paid even more."[17]

Leffingwell's bound journal ends with the September 3 entry. Similar to an earlier journal, in the final pages he recorded many lists and drawings. He also wrote trip summaries of three of his field experiences. And he detailed a number of lists pertaining to the *Argo*, including a list of its faults, a list of items needed, and a list of structural and deck changes to be made.

Winter Approaches

On September 4, Leffingwell and McIntyre headed west to continue mapping the coastline, outer reefs, and islands. They first had to restore signal locations and stations that had been blown down or washed away. Leffingwell was clearly pushing to get the work done before the arrival of winter. They reached the Flaxman base camp on the afternoon of September 21. They had been gone two and one-half weeks. Winds were consistently between 50 and 60 miles per hour the following day.

The month of October at the Flaxman Island base camp was relatively quiet. Leffingwell settled in by reading, cleaning, and worrying about the condition of the *Argo* in the wind and stormy seas. During the course of a week, they managed to winch the *Argo* higher onto the beach.

But by October 2, the ponds were frozen. At noon on October 3 Leffingwell looked out over the ice and counted 53 seals. Meanwhile, McIntyre worked on a new tent of silk, a new sledge, and made some root beer, which Leffingwell described as "fine." As the month progressed, Leffingwell hitched up the dogs to a sledge and hauled wood, water, and ice to place over windows and insulate the outer walls of their buildings. Temperatures outside fluctuated between lows of -7° and 0 °C. When the weather warmed up, he placed tents and tarps over the ice-covered structures and windows in an attempt to slow the melting. During the evenings Leffingwell observed the aurora borealis in impressive and bright displays of color.

They often had company: Arey, Wixra, and Gallagher and others. Indoors, Leffingwell made dog traces, filed knives, and worked on his instruments. He could not tolerate inactivity for very long, and by October 24, he had been reading and "loafing" for days and found life "monotonous."

As part of his baseline mapping, Leffingwell established azimuth marks in four directions from a center point near the house. The north azimuth marker had washed away, so on October 31, he built a fire over the location so he could set another marker. There was a full moon on November 5 and it was a beautiful evening. Life was slowing with the arrival of winter and the ocean was calm. Two days later, he wrote: "Winter at last. Fine clear day & cold...Went out to ocean am. Ice grinding about 100 yds from beach & pools of water & young ice...At noon Ned's [Arey] & Toklumena's outfits came over. Ice thin in places. Boys other day had to crawl on bellies to get ashore in dark from here. Ice broke same night west of here. Ned staid all night."[18]

November was quiet, with snow off and on. Leffingwell and McIntyre settled in and took a break from mapping. The temperature in the house reached 0° on November 10 for the first time. The Flaxman base camp hosted many visitors, including Arey and Gallagher, who worked on building new stoves. As for Leffingwell, he spent several days drawing up houses for the ranch property in Pasadena. It was a nice diversion from his work. Toward the end of the month he worked again on his triangulation plan for the coast.

While McIntyre was away retrieving some cached fossils, Leffingwell wrote about an unusual late evening occurrence:

> About 11pm heard apparently heavy crunching outside. Looked out of window for bear. Then quiet & I started to bed again, when crunching began. Dressed & went out with gun, around house, but didn't see anything. About 1pm heard it again. Hammered on walls & yelled & perhaps dog or fox. Didn't stop so dressed again & out—nothing. Slept rest of night in clothes gun handy & awakened often. This am found ermine? tracks on roof. had dug holes in snow all over & this is what I thought was crunching of bear.[19]

Most of December passed with Leffingwell attending to his accounts and other desk work. By December 10, he had finished a map at a scale of 1:80,000 that included soundings from Flaxman Island west to Oliktok. He wrote up and prepared a copy of the previous year's astronomical observations to mail out. Before he could start taking new observations, he spent several days working on securing the triangulation stations, and then he meticulously cleaned and adjusted his scientific equipment. Meanwhile, McIntyre and an Inupiaq helper hauled the engine from the *Argo* up to the house where the three of them overhauled it.

McIntyre baked bread and pies for the Christmas holidays. The house was full of company by December 23 and on December 25, they hosted a large dinner. Their guests stayed for several days.

As 1911 came to a close, Leffingwell had finished mapping a large portion of the coastline between Barrow and the Yukon border, and had mapped the geography and geology inland through the Sadlerochit Mountains to Lake Peters and Lake Schrader. The coming year would be busy, challenging, and productive in the mapping of the coastline. Wind and ice were two of the ongoing challenges, but there was also a question of whether Leffingwell would be able to leave for the lower 48.

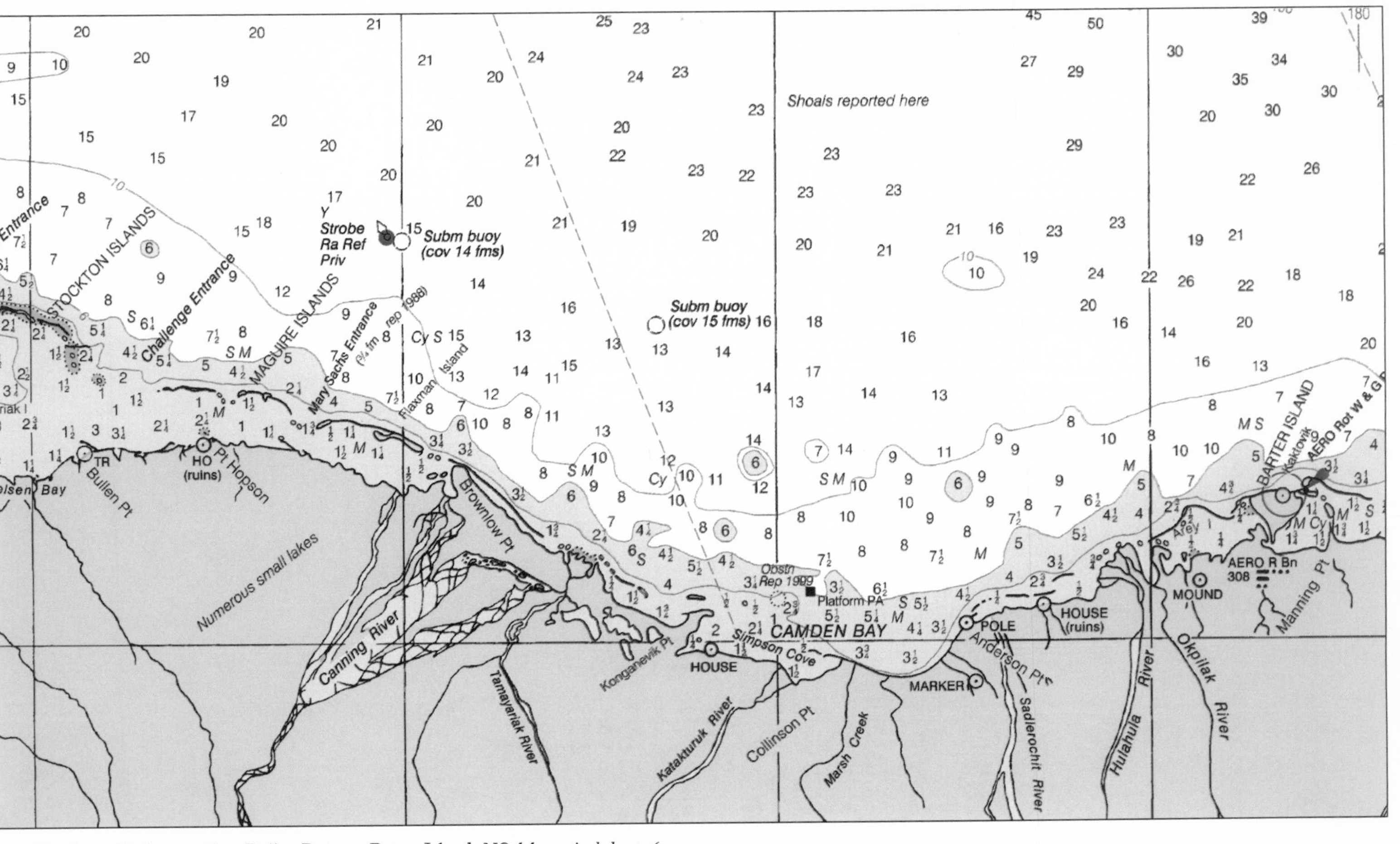

Northern Alaska coastline; Bullen Point to Barter Island. *NOAA nautical chart 16004*.

CHAPTER 10

Exploration of the Coastline Continues, 1912

Leffingwell began New Year 1912 by packing up six boxes of fossils, a drawing board, and box of paper. Now that winter had come, he turned his attention to building snow walls for his astronomical observatory and working with his transit telescope.

On January 6, he noted it was light enough to work for three hours without a lamp. He had problems with frost in and on the instruments, which necessitated that he bring them inside frequently to dry them out. Some of the instruments were not very stable and quite sensitive to environmental conditions. He wrapped them in cloth for protection. Even his breath affected the results. On one occasion he wrote: "Ice inside of lantern globe."

Temperatures were consistently in the negative 30s, sometimes dipping into the negative 40s, while inside the house they varied from 4° to -3°. Winds blew up to 50 miles per hour, yet Leffingwell was increasingly pleased with the quantity and quality of his observation results.

During February he worked outside when he could. He checked his Flaxman baseline net (or grid) for mapping by traveling to each marker, resetting if necessary, and placing new posts as required. The north, south, east, and west azimuth markers were located anywhere from one to three miles from the Flaxman base camp. From each point he eventually recorded a round of angles. During his travels, he often stopped by Arey's cabin for a visit and stayed the night. Leffingwell then organized and packed more items for future travel including the photography equipment, transit, chronometer, and theodolite. His mapping accuracy was high, within one second of a degree or approximately 100 feet.

West Again

Early March 1912 saw Leffingwell and McIntyre, four dogs, and two sledges headed west yet again. Leffingwell and McIntyre had brought along the latest version of their silk tent and Leffingwell wrote on March 8: "Cold air in around door. Stove in center of tent. Mack & I on either side. Single center stick & round (9 sided) tent, double silk. Light =11±lbs, but no place to hang gear. Much better than sqr [square] tent last year."[1] Before the end of the following week he had concluded that Inupiaq tents were far warmer.

They spent one night on Tigvariak Island before returning to camp on the mainland. It snowed early on March 16 and the following evening Leffingwell wrote:

> Blew up last night & tent flapping. Lay awake long time afraid it would carry away...Up 7:30 & had to cut door loose & dig out. Heaviest gale I have been camping in & drift also...blocked door. Put sleds up to windward of tent for break & built alleyway. Took two hours. Nearly got lost once 20 ft from tent in carrying snow block. Mack out noon to cut wood & had to go out thro [sic] top of alleyway...The tent is so deep in snow 18 in. below level that walls do not stand high enough... No more pole tents for me. Snug as a bug after alley way made, but roof flapping bothers me. Stove pipe steady as church steeple.[2]

The weather deteriorated so much over the next week that on March 23 they finally cached one of the sledges and almost everything else except sleeping bags, and took off for another of Arey's camps. Leffingwell calculated that they traveled 21 miles in seven and one-half hours. They reached the Flaxman house on March 24 and returned to Arey's the following day with a load of food, then returned to Flaxman Island.

The wind continued to blow between 40 and 60 miles per hour over the next several days and Leffingwell wrote several times that he was glad that they had returned home from the field work.

During the first week of April he took some photographs of the house and tent sticks using five-year-old plates to see if they were usable. He developed and printed the negatives with fine results. On April 4, the weather finally broke and visibility was good for the first time in several weeks. As spring advanced, conditions were bright enough that Leffingwell realized that he needed to start wearing snow glasses or goggles. Removing his protective goggles to use the instruments for even an hour caused his eyes to hurt. On the morning of April 10, Leffingwell and two Inupiaq helpers, Karoiga and Nanjek, along with three dogs headed west towards Bullen Point for more surveying.[3] McIntyre and

Arey accompanied them as far as Arey's place. The trio arrived at Bullen Point at 3:40 p.m. and Leffingwell later wrote: "Took a look around with glass while boys got cached sled. Fire one hour. Take willow sticks this trip & feel fine to be in round tent again. No door, loose cloth."[4]

The following morning they were off with the boat sledge under sail. Karoiga and Nanjek helped Leffingwell restore fallen signals and placed tripods so he could take angles for mapping. Temperatures were still in the negative teens and 20s, but warming.

Several days later he decided to return to Flaxman Island because he was concerned that the Sagavanirktok River would overflow its banks and leave them stranded. He wrote that there were "deep wide soft drifts" that made it difficult to see the route. They often sank to their knees but finally camped in the late afternoon on the west side of Prudhoe Bay.

When he traveled to Gull Island on May 7, he could not find the milk bottle he had buried below the surface in case the signal above had been lost. He described the scene: "Waves have been over it & washed nearly all wood off. The log old & 35 ft long was last summer 1/2 in sand. Now on S side sand gone...Put up small signal nearby. This makes this quad [quadrant] very weak."[5]

The following day was tough for travel and temperatures. They were traveling in a foot of soft snow, often breaking through the crust to sink another foot. Leffingwell did not believe the thermometer reading of -12°C, and wrote he was having more trouble in this setting than he had experienced at -40°C. It was cold enough that Karoiga slapped his hands to get feeling back into them. Leffingwell was surprised because he had never seen an Inupiaq do that before.

They camped at Heald Point on May 8 and at Anxiety Point the following day after traveling 10 miles. The three of them had become tired after only traveling two miles. They were running short of food. The cache at Stump Island was meager, not what Leffingwell had anticipated. He located one day's worth of flour and two days of food for the three of them, and no food for the dogs.

The next day before departing Leffingwell made copies of his work and they reached Tigvariak Island at 5:30 p.m. Later that evening Aigukuk and his party arrived. Toklumena provided Leffingwell with bear blubber for the dogs.

The following day, Leffingwell tried to hunt seal: "Saw same big seal, same hole as last...as K [Karoiga]...is sore, I tried it. Hands & knees, 150 yds then it got uneasy, so on belly for 250 yds behind drift 18 in high. Got tired & wet & had to lie on back to let heart quiet down. Seal didn't see me much & was quiet, but afraid to go closer, nothing for blind. Thot [*sic*] he was <150 yds...Built blind of snow blinds, but he did not show up again today."[6]

It was a good learning experience for Leffingwell. He arrived at Flaxman Island at 4:00 a.m. on May 15 and the Inupiat arrived later that afternoon.

A Visit from Stefansson

Leffingwell slept only three hours, then spent the day cleaning up and sorting gear. He gave the Inupiaq assistants some tools and other items that he no longer needed. The wind blew up to 50 miles per hour at night but by the following afternoon it was calm, clear, and felt warm at 0°. While Leffingwell made a new plane table for mapping, McIntyre and the Inupiat hauled the boat up on a new sledge McIntyre had built.

On May 17 it was evident that spring had arrived, and he wrote: "Over to N mark & banked it up with snow...Dug trench thro drift in front of house to drain away from door. Opened summer door. Mack fixing boat. Busy all day, but not much done. Mack's dogs (3) on roof kept me awake...Got up 4:30 & threw sticks at them. Worked from there on. Snow hard this am softened about 10 am. Sweet spring for sure. Water sky in patches far North."[7]

Stefansson, accompanied by an Inupiaq, arrived on May 18 from Herald Island, and Leffingwell and Stefansson spent most of the next few days talking. By May 21, Leffingwell said his voice was nearly gone but that Stefansson was "still fresh." Stefansson left about 11:00 p.m. that evening. During one of their breaks from talking Leffingwell read a letter from home that Stefansson had delivered and wrote letters to send out.

After Stefansson's departure, Leffingwell spent the following days working with McIntyre on the *Argo's* engine, which seldom worked properly. He also worked on packing gear that would be sent south later in the summer, and then visited with Arey, when he and Gallagher arrived on May 25. Leffingwell was hoping to complete his work in 1912 and head south. Karoiga had sent Leffingwell a note that read, "Plenty water all ice." He could not leave yet but was ready. The details he required in his mapping were covered by ice so in the meantime, he packed the books that would return south with him.[8] By June 2, Leffingwell was getting impatient about not being able to travel and complete more mapping of the coastline east of Flaxman Island.

East Along the Coastline

Finally, in late afternoon on June 6, after being at the Flaxman Island house since May 15, Leffingwell and two Inupiaq assistants departed for Brownlow Point with the boat sledge and a three-foot trailer. They arrived at 9:00 p.m. and continued on to the east end of the spit and established camp just before midnight. The next two days were spent mapping as they traveled first to Konganevik Point and then on to Collinson Point. They stayed four nights at Collinson Point. Leffingwell had not been feeling well, had no energy, and was sick with what he called "grippe" (the flu). According to his journal of June 12, he had slept only twice in five days. He was exhausted and finally caught up on some much needed sleep.

His energy restored, they moved camp the following day to the spit of the Sadlerochit River. From the camp Leffingwell explored an area inland that he had tried to reach before. He found more fossils and a new geologic formation with good exposure, which he believed was of Late Tertiary age, between 38 million to 1.6 million years ago.

While Leffingwell mapped, the others tried to move camp using the sledge to move the boat. Leffingwell ended up helping them and they arrived at the next campsite about 10:00 p.m. He described it as a wet place with wood, but no sand. They pitched their tents on the tundra. At midnight Leffingwell went to retrieve the plane table he had left at the mapping site and fell as he tried to jump the creek. In the process he broke the tripod. He arrived back at camp in the fog at 3:00 a.m.

By the end of the following day, Leffingwell and the boys were camped west of the Hulahula River flats. Traveling became even more difficult as they moved east. When the water along the shore disappeared under ice, they had to haul the boat and sledge over the ice. They dumped part of the load on shore and sledged the boat overland to their next camp, which they reached about midnight.

It was hard work and two days later Leffingwell described how they grounded 50 yards from a beach on Arey Island, poled slowly against the wind, and occasionally got out of the boat to shove it along. When they finally tied up they were still 20 yards from the beach of the island. It had taken an hour to travel the quarter-mile distance and his knees were "cold and stiff."

When the weather turned to rain, they remained in camp. They could not use the fishing net because the drifting ice would likely destroy it or carry it away. Leffingwell had already mapped the island in 1907, so he did not have much to do other than wait for the ice to move out so they could travel again. It rained hard for a couple of days, the winds blew at more than 20 miles per hour, and he spent much of his time in the tent. He got out and walked when the weather temporarily let up.

The winds calmed and June 24 looked promising. Leffingwell saw a hazy sun and cumulus clouds over the Sadlerochit Mountains. They finally departed Arey Island at 4:30 a.m. and headed toward Barter Island, traveling 17 miles in 10 hours. Their travel alternated between poling in the boat to loading the boat on the sledge and hauling it across ice. At times they traveled less than a mile in an hour's time. He and the Inupiat continued the difficult work until the sledge carrying the boat broke through the ice. They took the boat off the sledge and proceeded to drag it 40 yards to water once again. They reached camp on the mainland southwest of Barter Island at 2:30 p.m. Not surprisingly, Leffingwell wrote that he was tired and turned in at 7:30 p.m. He took the following day off and caught up on sleep before he turned his attention to taking angles. One of the assistants made a map of the coastline west of their camp. They stayed five nights at the camp during which it snowed and the winds blew at 30 miles per hour. The previous year they had rarely used the stove in the tent to keep warm, but in 1912 the stove burned continuously. Leffingwell predicted that it would be a long time before the sea ice drifted out.

Mapping Barter Island and Environs

Their next camp was located on the south side of Barter Island. Over the next five days, Leffingwell established several additional triangulation stations with the help of one of his assistants. His appreciation was duly noted: "K has put up fine signal on peninsula."[9] Ice was still plentiful and Leffingwell concluded that they might have to use the sledge in the near future.

The weather was sunny July 1. In late afternoon Leffingwell set off to map the west side of the island and most of the north side. By 10:00 p.m. the fog rolled in and he wrote that his hands were "so cold had to slap them." When Leffingwell started for camp in the fog he became disoriented. He saw a signal that they had placed two days earlier on June 29 and knew he was headed the wrong way. In his typical substantive recounting of circumstances he wrote: "Then curved around that to NW & came out near NW corner Id. Knew that wind was on wrong cheek & looked at compass & checked wind. In breaks in fog saw that I was keeping signal abeam to right yet thot [sic] I recognized ponds etc in fog & was 70° off course."[10]

He arrived back at camp at midnight, had supper, and finally turned in at 1:30 a.m. Between Leffingwell and his two assistants, they had polished off the first sack of rice and flour since leaving Flaxman almost a month earlier. And Leffingwell had calculated that they were using two pounds of butter a week on their pancakes.

Leffingwell mapped the east side of Barter Island then headed south toward what is known today as Manning Point. They established camp where the spit connected with the mainland and stayed three nights. The weather was calm but the ice was no good for sledging, so they stayed put.

The following day, he mapped the mainland coast south of Barter Island for about three miles to the east, ending close to the mudflats of the Jago River. He finished the work within a few hours and returned to camp, noting that most of the shoreline was clear of ice. He hoped the ice was moving out so he could keep mapping. As they continued to their next camp, he thought about future plans. He hoped to catch a whaling ship headed south, but was having his doubts. He concluded that the ships would probably overwinter.

Fog and winds that varied from 35 to 60 miles per hour forced them to stay six nights. He had told McIntyre that he planned to return to Flaxman on July 9, but instead they were now camped on the east end of what was probably Bernard Spit. Over the three weeks in the Barter Island area, Leffingwell wrote that he had managed only five short days of work. He focused on the wind's effect on the movement of the ice, and hoped for a strong west wind to blow the ice out. By July 12, conditions remained windy and he was tent-bound drawing houses for his father's Pasadena, California, property. It was a pleasant diversion.

Return to Flaxman Island

The next day, Friday, July 13, Leffingwell decided to return to Flaxman Island. By July 16, after traveling nearly 12 miles in seven hours, Leffingwell and his assistants reached a former camp located west of the Sadlerochit River mudflats. There were three other tents at the camp and they stopped long enough for a meal of fish before continuing west and making camp. Leffingwell caught two fish later in the day and by the following day he and the boys had cleaned and racked about 20 more.

Leffingwell mapped a couple of miles then watched the Inupiat for hours set the nets for fish before returning to camp. The trio had caught two seals, so with hearty appetites, they ate one for dinner and one for breakfast. Leffingwell wrote that he was hungry for fat. One of the Inupiaq went to O'Connor's place at Collinson Point to get some flour and returned with a supply of flour, sugar, lard, tea, and tobacco. Later that afternoon they moved camp toward Collinson Point. Some of the distance was made in open water, some as they chopped their way through ice. But by the next morning the ice had moved in behind them and was up on the beach they had departed. Based on the delay in their return to Flaxman, Leffingwell surmised that Arey and McIntyre were probably worried about him.

When the winds changed, the movement of ice did as well. It would clear out a bit, then close back in. He and the boys moved camp when they could, even if just a mile. By late afternoon on July 24, Leffingwell had mapped to Collinson Point and stopped by O'Connor's place for food and magazines. Upon his return to camp he explored a bit up Carter Creek, finding some fossilized wood.

Late the following evening, when it became apparent that the only thing moving into the area was rain, Leffingwell and his Inupiaq assistants decided to pitch a tent. Their eighteenth camp was located a mile east of Marsh Creek. As it rained, the ice moved out enough that they could have reached Marsh Creek. However, rain, wind, and rough seas kept them tent-bound through the next day.

Finally on July 28, the seas calmed considerably and, in a change of plans at Marsh Creek, Inupiat joined those in the boat, now eight in total. Leffingwell left five dogs with one of the Inupiaq assistants, and with the others traveled 10 miles in five hours and camped south of

Konganevik Point. Toward the end of their travel that day, they poled the boat until the ice blocked their progress. Leffingwell wrote that he was sorry he had not finished mapping the end of Collinson Spit recently. It was not going to happen this trip.

Leffingwell saw water near Konganevik Point, but beyond that there was ice as far as he could see. There was a half sack of flour remaining and a small amount of tea. Based on the ice conditions and low food supply, he decided to walk to Flaxman Island, a direct distance of about 21 miles, and have one of the Inupiat bring the boat when the ice cleared. Leffingwell left late that afternoon carrying roughly 20 pounds of gear and traveled about five miles upstream before he could cross the creek. He crossed another small stream, believing that it was the east branch of the Canning, but at 4:00 a.m. located the real one, and it was in flood. Leffingwell was tired so he stopped to cook some meat. His energy restored, he reassessed the crossing and knew that it meant getting his books wet and going for a swim. He wisely decided against it and started his return to the camp he had left the afternoon before. When he arrived back at camp, no one was there.

The ice had moved out and the rest of his party had moved on. Leffingwell could not spot a tent in the distance, so he figured they were west of Konganevik Point. He built a smoky fire, hoping that the Inupiat would see it, and dried out his foot gear and ate a meal. Leffingwell decided he would have to continue the eight or nine miles to Collinson Point.

At Collinson Point, Leffingwell slept on the floor at Duffy O'Connor's place. Based on his pedometer, he calculated that he had walked 58 miles in less than 36 hours. "My best performance," he wrote, "but don't want any more of it."[11] His feet were sore but he felt he could have gone further if he hadn't been carrying a pack. Much of the next day, August 1, was spent talking and visiting. Late that afternoon he sighted a two-masted schooner out several miles and traveling east. It was McIntyre and Arey on the *Argo*, accompanied by Gallagher and Karoiga. They arrived on shore a little after 6:30 p.m., departed with Leffingwell by 7:00 p.m., and sailed west until midnight. The Inupiat had left the camp shortly after Leffingwell and made it to Flaxman the following day. When McIntyre and Arey learned that Leffingwell was not with them, they set out in the boat and sent Gallagher and Karoiga back out to look for Leffingwell along the Canning River.

Flaxman House

They all arrived back at the Flaxman base camp on the afternoon of August 3, after having been away since June 6. There were 15 to 16 tents pitched at Brownlow Point. Leffingwell settled his accounts with his Inupiaq helpers, then Arey and McIntyre. He had not slept much in the last day and a half so turned in early. The following morning he immediately organized his gear and resumed work on the *Argo*'s engine. Gallagher worked on it the following day and Arey and McIntyre moved on to the *Argo*. In lieu of wages, Leffingwell gave the *Argo* to McIntyre, who planned to sail east. After Leffingwell took an inventory of the provisions, he packed and stored gear.

By August 6 Leffingwell had stored his instruments and other valuable gear, and knew he would be returning north once again. He buried the house keys in a metal tin and waited for the fog to clear, the winds to die down, and the seas to calm. He had spent much leisure time drawing ships, but after observing recent gales and rough seas, was happy to be ashore. He scrapped his idea of building a boat.

To Barrow

Leffingwell was anxious to complete more triangulation work before traveling west to Barrow, where he hoped to catch a southbound whaling ship. The following night he wrote that he needed to be leaving "pretty soon." He pulled together food and gear for the trip to Barrow, including fox and caribou skins, presumably to use to pay off accounts or sell for cash. With a group of Inupiat, he finally began the journey by canoe to Barrow in the morning of August 15. They were so loaded down that Leffingwell feared the skins would get soaked. While at Tigvariak, six more canoes arrived with Inupiat who were amenable to carrying some of his load; the group would travel together.

What was supposed to have been just a few days at Tigvariak turned into almost two weeks of fog, wind, and occasional rain. Stopping at the island were canoes and whale boats, coming and going both east and west. Strong winds kept the canoes on shore. There was nothing Leffingwell could do. The *Argo* was off to the east and could not help.

With each arriving boat came news. Leffingwell learned from the Herschel Island Inuit (Inupiat in Alaska, Inuit in Canada) that there were men from the Boundary Survey at the coast. The Boundary Survey was part of the joint Boundary Commission created by the governments of the United States and United Kingdom, charged with locating and marking the Alaska-Yukon Border. Their presence was a significant statement of the increasing importance of the region in commerce and exploration.

On August 25 Leffingwell spent another day reading; it was his last book, and he finished it the following day. Leffingwell was skeptical that he would be able to rendezvous with Captain Cottle at Barrow. If he did not meet Cottle it likely meant another winter in northeastern Alaska. Two days later he wrote that much ice remained around the islands and he hoped that the ice pack was delaying Cottle.

At 7:00 p.m. on August 28, an Inupiaq named Manixrak and others returned in a whale boat and loaded Leffingwell and all of his gear to head for Barrow. They departed at 8:00 p.m., fog immediately set in, but they continued on west. When the seas became too rough to travel, they cut to the inside of the islands and took soundings when they needed to. They passed Heald Point, Prudhoe Bay, before reaching the east end of Stump Island at 1:30 a.m. They took the sail down and had to pole for a half hour in the shallow water. But it was better than traveling to the outside of the islands and rough seas. They had also stayed to the shallow waters to avoid running into ice. At 4:00 a.m., cold and wet, they decided to camp on Cottle Island.

On August 31, they departed their next camp on Amandliktok Island at 5:45 a.m. It was foggy and Manixrak decided to travel the inside route. It was tricky; Leffingwell took soundings every five minutes and was "glad to get" them. They reached camp near Cape Halkett late in the afternoon and Leffingwell went to sleep without cooking a meal. Their departure the next day was not until shortly after 11:00 a.m. and in late afternoon they encountered a whale boat traveling east. Leffingwell learned that Cottle had not yet reached Barrow, so there was still a possibility that he would have transportation south. They camped west of Pitt Point that night and left the camp the following morning accompanied by favorable winds.

Late afternoon on September 2 they at last reached Barrow. Leffingwell was much relieved to learn that Cottle had not yet arrived. He slept at Hobson's house where he had stayed previously, and had breakfast the following morning at Brower's place. He spent the day reading a large pile of mail, and fixed a phonograph and a camera for a couple of the local residents.

Leffingwell's journal daily entries ended with September 3, but as before the pages at the back of the journal contained lists, notes, and drawings. They included a game list for July; a list of items for McIntyre in 1914; a list of items to bring up; notes about a conversation with Arey regarding bowhead whales; a list detailing bird counts, description, size, sketches; bear info; a task list; names and addresses; and miscellaneous notes. He apparently did not keep a journal during his journey south from Barrow as he had done in 1908.

Leffingwell had hoped to complete his work in 1912, but was unable to do so. He would need to return for his equipment and complete his fieldwork. Even with the help of the Inupiat, the work had been hampered by weather and food and shelter requirements. His years in the Arctic between 1909 and 1912 were summed up best by an Iowa newspaper article of November 9, 1912, headlined: "Survey in Arctic Regions No Job for Man in Hurry, Says Ernest Leffingwell."[12]

Leffingwell had accomplished much, but if he was to be comfortable with the quality of his work, he knew he would have to return north and finalize it. Given his experiences on the *Argo* and the *Duchess of Bedford*, it was unlikely that he would choose to procure another boat. Stefansson, as commander of the Canadian Arctic Expedition, would provide him the opportunity to return north in 1913. Although Leffingwell was tied only peripherally to the expedition, its outcome would be tragic and controversial for decades.

CHAPTER 11

Return North on the *Mary Sachs*, 1913-1914

When Leffingwell departed Alaska in fall 1912, he knew he would have to return the following year to finish coastal mapping and retrieve his instruments.

As it turned out, 1913 was a challenging year on the north coast of Alaska. Reported to be one of the worst seasons ever seen, the presence of pack ice provided a very narrow window for travel. Whaling ships found safe harbors and overwintered. Thus, Leffingwell was forced to sail from Seattle on June 24, 1913, aboard the passenger ship, *The Senator*, bound for Nome where he would meet Stefansson and the Canadian Arctic Expedition (CAE). Some of the other expedition members also traveled on *The Senator*.

Leffingwell wrote: "In 1913 a Canadian expedition was sent out under Stefansson...and Anderson to cover practically the same ground that I had, and I went with them to act as ice pilot, as I wanted to bring back my valuable instruments."[1]

On May 29, 1913, the Department of Naval Service wrote Stefansson and defined the two primary goals of the CAE: "exploration of unknown seas and lands, and second, the gathering of scientific information with respect to these

Supply ship *Mary Sachs* at Peard Bay on the North Slope, 1913. *USGS Photographic Library, Leffingwell Collection* areas, and also to the partly known lands and seas in the vicinity of Coronation Gulf."[2]

The Canadian Arctic Expedition under Vilhjalmur Stefansson, on board the *Karluk,* departed Victoria, British Columbia, to northern Alaska and Canada on June 17, 1913.

The CAE was initially supported financially by the American Museum of Natural History, Harvard Travellers' Club, and the National Geographic Society. Requiring additional funds, Stefansson solicited and received major financial support from the Canadian government, hence the expedition name.

The *Alaska* had been purchased as a supply ship for the *Karluk.* When Stefansson reached Nome on July 8, he was given approval by the Naval Service to procure an additional supply ship, the *Mary Sachs,* 56 feet in length and built in 1898. The *Karluk* and the *Alaska* were simply not large enough to contain the expedition's provisions and equipment. The plan was to divide the expedition into two parties: a Northern Party to conduct scientific studies in northern Canada, and a Southern Party to explore the Beaufort Sea north of Alaska and western Canada with the *Mary Sachs* at their disposal.

Leffingwell's connection to the CAE was peripheral because he needed transportation to return to the North Slope of Alaska. But his role became important to the success of the expedition in navigating the northern coastline. Stefansson felt that Leffingwell's map and knowledge of the coast was the best available. Captain Bernard of the *Mary Sachs* had not been to the northern coast of Alaska, so Leffingwell would assist him with navigation. It was a mutually beneficial arrangement. Leffingwell had not forgotten that he was the one that invited Stefansson to join the Anglo-American Polar Expedition in 1908, Stefansson's first trip to the Arctic.

Over his years on the North Slope, Leffingwell gained experience with different types of boats as he traveled along the northern coast, and later described them in his report. A canoe was too easily swamped by waves. The heavy, 30-foot whale boat he owned for a brief time was unsuitable for his working environment. He appreciated the oomiak, also about 30 feet in length, for its light weight, its shallow draft, its carrying capacity, and the ease with which one or more men could transport it. What Leffingwell found missing was a centerboard or keel to help travel against the wind. He finally purchased a 27-foot, flat-bottomed dory that met his needs. It was lightweight, had a large carrying capacity, was easy to beach, and could travel the easiest of any craft against a wind. Leffingwell and the dory went aboard the *Mary Sachs* in Nome, with Captain Pete Bernard. Much later, in January 1914, Stefansson wrote: "Capt. [Bernard] tells me that he [Leffingwell] did more work than any man on the ship going up, taking his turn at everything and doing many others out of turn."[3]

The *Alaska* remained in Nome for repairs but according to Leffingwell's journals, the *Karluk* left Nome on July 22. The *Mary Sachs* departed the same day with Captain Bernard and Kenneth Chipman, topographer, in charge. In addition to Leffingwell, on board were a sailor, an engineer, and an Inupiaq man, woman, and baby. Plus 11 dogs. Although Leffingwell was miserably seasick, he provided invaluable guidance to Bernard.

The *Karluk* and *Mary Sachs* parted ways near Cape Prince of Wales. As the *Mary Sachs* reached the Seahorse Islands north of Wainwright, Leffingwell wrote, "Ice all heavy ahead." They spent 11 days there and made a mere 20 miles, carefully negotiating grounded ice and shallow waters closer to shore. Leffingwell was careful and conservative in his recommendations. Not everyone appreciated his caution and experience, and some crew members were quite vocal in their disagreement with his approach. At one point, Leffingwell offered to step down as pilot through the area, but Captain Bernard declined his offer. It would later become clear that Bernard's decision to retain Leffingwell was wise and justified.

When they finally reached Barrow August 15, Leffingwell learned that no ships had arrived from the east because of the extensive pack ice. As became evident later that year, the tragic outcome of the *Karluk* was due in part to the pack ice conditions.[4] The *Mary Sachs* departed Barrow that same evening.

It was one thing to deal with extensive ice and shallow waters. But navigation was made more difficult by the fragile nature of their instruments, as evident on August 17, when Chipman wrote that they had been using Leffingwell's compass and at some point, it had been moved three feet. The result was that tools in the engine room had thrown off their location readings. Chipman wrote: "Last night in crossing Harrison bay I sounded four fathoms and very soon after got only seven feet.

We pulled out quick!…we are mighty glad to have with us Leffingwell who knows every mile of this coast."[5]

By that evening they had anchored at Point Mack and Aigukuk dropped by for a visit. Aigukuk told Leffingwell that he had seen smoke from the *Karluk* and that it was far from land. During the night, the weather cleared off and they continued their journey, arriving at Flaxman Island about noon on August 18.

At the time of his return to northeastern Alaska, Leffingwell believed that he would finish his work, collect his instruments, and leave by the end of summer. But heavy ice conditions eliminated any possible return south. Leffingwell had basic provisions, but no sledges or dogs. He would spend six weeks alone and the isolation would take a toll.

Flaxman Island Base Camp

When Leffingwell arrived at the Flaxman house, all was well. He observed that weather conditions were about a month ahead of usual. There was little water at the ocean, mostly ice, there was snow on the ground, and temperatures low enough that ice had formed on the pond.

Leffingwell took inventory of his provisions, both clothing and food. He had left some clothing at the house, but would need a few items including a sleeping bag. He determined that he had adequate food except for butter and bacon. Leffingwell purchased 45 brant (small geese) from Terigloo and Natkusiak, then traded 25 of them to Captain Bernard for kerosene and pork. His daily diet was simple: pancakes for breakfast, vegetarian meals or caribou and eggs for supper.

He returned to the *Mary Sachs* anchored nearby on August 25 and ended up staying the night. The fog was heavy and snow fell all night. The following day they put Leffingwell ashore, where he quickly mapped the northern edge of the island west of Flaxman.[6] He was then dropped off at the west end of the Flaxman Island spit before noon and he returned to the house. He spent the afternoon gathering and piling wood.

In late August Leffingwell settled back into the house by cleaning and writing letters. Over the course of a week, temperatures dropped, there were gale-force winds, drifting snow, and Leffingwell kept looking for ships, primarily because he wanted to send some of his instruments home. He repaired the shed roof with pieces of canvas, sail, tarp, and burlap. The house roof could not be repapered because of the cold temperatures, but he did manage to place sod where needed. He turned his attention indoors and spent the first week of September working on his triangulation calculations from the previous year. It was a difficult task because he did not have his original notes, likely because he planned to return that fall. Or perhaps he forgot them, or felt he could accomplish what he needed without them. But it was a good week to be working indoors as the wind was blowing up to 30 miles per hour.

On the afternoon of September 6, he sighted the *Alaska* at the northwestern end of the island and quickly went to meet it. The *Alaska* had reached Barrow on August 19 and found heavy ice from Barrow to the Colville River. Leffingwell noted that they had used his map of the lagoon, and had not seen any ships headed west. One of the men who came ashore was O'Neill, geologist with the CAE, to return his map of the coastline. O'Neill wrote that Leffingwell had told him about the ponds freezing over a month early and provided them with sketches of places to overwinter if they did not reach Herschel Island. O'Neill also observed a few cliffs where previously melting had undercut the banks and blocks of land had fallen away. Leffingwell gave his outgoing mail to the *Alaska,* and she continued east.

Chipman was quick to recognize the quality and effort of Leffingwell's work. He wrote a letter to his superior at the Canadian Geological Survey on September 7, 1913, detailing Leffingwell's methodology, equipment used, and willingness to share information: "His work is good and has been of great value to us...I was very glad to have some four weeks association with him."[7]

Chipman also wrote that Leffingwell was not fond of Stefansson's methods. Leffingwell was meticulous and organized in conducting his fieldwork. Stefansson appeared his polar opposite. At Nome, provisions aboard the *Karluk* and the two supply ships were disorganized and had divided the men from their equipment. Expedition members later determined that equipment including traps, picks, and shovels needed by the Southern Party was aboard the *Karluk*, never to be seen again. The *Karluk* was to become trapped in the ice off of northern Alaska while sailing to Herschel Island. The drifting pack ice would carry the ship close to Wrangel Island north of Siberia where the ship was destroyed by the ice and sank. Eleven crew members died in the aftermath. Stefansson had left the ship with companions before it started drifting and shortly after it was caught in the pack ice off the north coast of Alaska.

One night Leffingwell wrote that he still saw light to the north at midnight and familiar stars. Now that he was alone, the familiarity must have been comforting. It seemed that he was always dreaming of other places and planning logistics. Over the next two days, he wrote that he was working on an "outfit" for a trip to Great Bear Lake in the northern Northwest Territories. The trip was never made and there is no mention of it elsewhere in his journals.

On September 9, a whale boat arrived from the east with a Point Hope resident, Jim Allen, accompanied by several Inupiat. He brought news that the *Mary Sachs* and *Alaska* were overwintering at Collinson Point; travel east was impossible. Other ships were frozen in near Demarcation Point. The *Karluk* had not been seen and Allen had no news from Herschel Island. The group stayed the night and Leffingwell traded items including a gun, ammunition, and lamp for other ammunition and bacon. They left early the following morning and Leffingwell provided Allen with his outgoing mail and a chart to Point Hope.

Over the next few days, Leffingwell bent some wood runners for the sledge, closed off his bedroom from the main part of the house to keep it warmer, and worked on the triangulation network. On September 14, Leffingwell walked to the ocean on the north side of the island and found an ice choked ocean. No ships would be passing. The following day he observed about 100 loons in several flocks heading for warmer destinations. He noted that most of the gulls had left and there were very few shorebirds remaining.

Preparation for Winter

Temperatures dropped and ice increased and on September 17, Leffingwell cut six-inch-thick ice slabs from the pond for use in front of the shed as insulation and protection against the harsh winter to come. Handling the slabs was not easy work for one person but he managed. Several days later, he washed clothes and scrubbed out his room. He finished building a six-foot, 32-pound sledge. Winds blew up to 50 miles per hour the following day and the house and windows flexed from the force of the winds.

The wind was followed by rain, sleet, and snow. Leffingwell spent much of his time working on the triangulation network for his mapping of the coastline, often working late into the evening. By the last week in September temperatures were generally below freezing, so he pulled out his winter clothing. It snowed about three inches on September 28. When he ventured outside, he hauled wood and water, hauled slabs of ice to the storage shed, or banked snow against the house and on the roof for insulation. During the day he walked to the ocean and observed a lane of water with ice moving toward the east at a slow rate. During the evenings it was common for Leffingwell to occupy himself by drawing and planning tents, houses, and boats. A few of the drawings appear in the back of his journals. It was a lonely existence.

Mapping Resumes

Leffingwell traveled south to the mainland on October 2 for a day of mapping. It was his first physical exercise in some time. Two days later he wrote: "A perfect day. calm & clear. Hardly a breath of air all day. Hot in sun; had to put shade up to keep ice front of shed from melting. Squared off a piece of Duchess boom & hauled it a mile and set it up as signal & observing post for NW Δ. [triangulation station] Back noon. pm overhauled theodolite."[8]

It was foggy all day on October 5. Captain Bernard and O'Neill arrived from Collinson Point for a two-day visit, and they talked more than slept. Bernard hauled wood and water while Leffingwell cooked mush for Bernard's dogs. Bernard and O'Neill departed in the early morning of October 7. Leffingwell traveled with them for a short distance, then parted ways to hunt for ptarmigan.

For the remainder of October Leffingwell kept busy with triangulation field work and spent several nights at Aigukuk's igloo west of Flaxman Island. When he returned to he made sealskin moccasins that covered his fur kamiks. On October 19 the wind was strong enough that Leffingwell observed the rafters shaking and house bulging. He recorded temperatures at -23° and -18°. Leffingwell kept busy sawing wood, building shelves for his desk, and building an air shaft from the old house to the new. Another gale blew on October 30. If Leffingwell had not been concerned about the previous gale, this one caught his attention. He wrote that the house was shaking like never before though he recorded no damage.

As October came to a close and daylight hours diminished, Leffingwell needed a lamp between 3:30 p.m. and 8:00 a.m. Leffingwell recorded his use of kerosene during September and October, as always maintaining precise information about anything and everything. His survival depended upon it. Monitoring his use of kerosene was critical, especially with winter approaching.

For the men of the CAE in safe harbor at Collinson Point, their thoughts were with the expedition members aboard the *Karluk*. Chipman wrote of a conversation he had with Captain Cottle who felt that the *Karluk* was gone. Captain Cottle's experience in the Arctic spanned 26 years and he had expressed that it was the worst year he had ever encountered with ice.

The members of the Southern Party were settling in to winter and trying to find their way. Leffingwell had previously shared his challenges in defining and mapping the geology of an area probably within the Sadlerochit Moutains with O'Neill. O'Neill wrote his superior at the Canadian Geological Survey: "I can't see where I could do much back here this winter along our line, although Leffingwell suggested a complex area back in the mountains that had him guessing but I don't imagine that working out fault systems would be very pleasant here in the winter: we have had a taste of it already with 8.5 below zero yesterday."[9]

In November, Leffingwell was still banking snow against the house, which significantly helped to maintain warmer indoor temperatures during big winds. He was still outside taking angles from his baseline triangulation stations, but it was cold and windy and he was not satisfied with the results. By mid-November, with only about four hours of daylight, Leffingwell decided to stop his outside work until winter had passed. In the mornings, he found ice in his wash basin.

Leffingwell's routine varied little; he always had projects. He spent time reading, performed calculations of his triangulation network, planned a new theodolite, made a new plane table for mapping, and worked on a tripod for use in the mountains. He cleverly designed the tripod legs to fold so he could use it as a walking stick. On more than one occasion, Leffingwell walked north onto the ocean ice and watched the ice form pressure ridges. It was great exercise and on his first trip he walked about eight miles. He hauled more ice slabs to place in front of the shed, and decided to make a canvas door for the shed to prevent the inevitable accumulation of snow inside.

Stefansson Visits

On December 1, Aigukuk and Toklumena arrived for the evening. They brought a letter from Collinson Point that the *Karluk* had not reached Herschel Island. The letter also indicated that Leffingwell's *Argo* was east of the Mackenzie River. Two additional ships, the *Belvedere* and *Polar Bear* were frozen in near the Aichilik River just west of Demarcation Bay.

On December 8, he wrote: "Stefansson, McConnell & Wilkins came in from west about noon. *Karluk* frozen in abreast of here about 20 miles out to sea on Aug 17 & drifted westward to Oliktok where S [Stefansson] & party left her for hunting. In gale she went away to west & they followed to Barrow. Natives reported seeing her near Smith Bay but nothing reliable since. They went to Barrow & sent out mail & are now on way East. 18 dogs all very hungry & thin. Talked till 1:30am."[10]

Leffingwell and Stefansson talked so much that Leffingwell had a sore throat two days later on December 10. They spent late nights and early mornings deep in conversation. Whether he recognized it or not, Leffingwell had been struggling with the lack of human contact. McConnell wrote that Leffingwell was unaware of the *Karluk*'s situation and thought that Stefansson and the others were visiting for Christmas. McConnell and Wilkins turned in about midnight while Stefansson and Leffingwell talked into the early hours of the morning. McConnell wrote that everyone talked so much they forgot to eat and that the number of meals consumed were diminishing as the days passed.

Over the next couple of days McConnell and the others expressed concern about Leffingwell. McConnell went so far as to suggest in his journal, "were he in civilization I would say he is on the verge of a nervous breakdown."[11] It was readily apparent that the isolation had affected him. McConnell wrote the following day that Leffingwell was: "so nervous that he can hardly sit still. He burns about a hundred matches a day, as he never takes more than a half a dozen puffs at his pipe after lighting it and, of course, when he begins talking again, the pipe goes out."[12]

By the time Stefansson and the others left early on December 14, Leffingwell acknowledged in his journal that he was having trouble enunciating. The unexpected visit by Stefansson, Wilkins, and McConnell had affected his mental health in a positive way, but also seriously impacted his food supply. Over the week's time, the visitors used one and one-half sacks of flour, which was one-third of Leffingwell's remaining supply. The flour would have lasted him four months and the visitors also consumed his supply of meat. As it was, he finished his supply of potatoes the following day.

Leffingwell was exhausted and within 30 minutes of his guests' departure, took a nap and had a dream he found unsettling: "I was in this house but in a strange room in which many forgotten objects of long ago appeared. One of Stef's men entering & asking me whether I knew whether I was on foot or horseback, I replied—That I was under a hallucination...but knew that I had the will to come out of it at my pleasure. Immediately I woke up & was somewhat frightened before becoming fully awake."[13]

Stefansson later wrote about a 1913 visit to the house at Flaxman Island and how Leffingwell's perspectives about necessities had changed over the years. Stefansson wrote: "The house had been added to and was rather palatial for those latitudes. He had an extensive library in several languages, one of his rooms was furnished with a roll-top desk, and altogether the equipment ranged from the sumptuous almost to the effete."[14]

He then wrote that Leffingwell had been a "tenderfoot" when he first arrived with many creature comforts from home that included a variety of jams, cereals, and other food. Stefansson observed that throughout his years in northeastern Alaska, Leffingwell had simplified his needs, including the foods he ate.

Christmas with the CAE

Leffingwell had been invited to spend Christmas with the Southern Party of the CAE at Collinson Point, where the supply ships *Mary Sachs* and *Alaska* were overwintering. He may or may not have recognized the impact isolation had on him, but the men of the CAE did. In preparation for the trip and for practice, Leffingwell built a snow house in case he had to sleep out once he reached Collinson Point. Then, on December 19, Aigukuk and Kopak's families arrived from the west and stayed with Leffingwell for a few days. That night the stove pipe burnt through, a repeat of his experience in 1908, except that this time he had his clothes on. With some help from Aigukuk, who stuffed snow down the chimney, and Kopak, who brought in snow, Leffingwell was able to cool the stove pipe, preventing the house from burning down. It took him several days to repair the pipe and in the meantime, the house temperatures dipped as low as -17°. When he tried to repair the stovepipe on top of the house, the weather did not cooperate. Winds blew over 50 miles per hour for several days and Leffingwell wrote that it was hard to keep from being blown off the roof.

Leffingwell was to leave for Collinson Point on December 22 and meet Captain Bernard halfway, a distance of about 17 miles, but delayed his departure due to 50 miles per hour winds. He waited until the next day when Aigukuk and Kopak decided to head for Collinson Point. They left at 11:30 a.m. and that afternoon met McConnell of the CAE and an Inupiaq known as Fred, who had been sent to accompany Leffingwell. They arrived at Collinson Point the following afternoon and McConnell wrote that Leffingwell was much quieter than when they had visited him at Flaxman.

Leffingwell's entry for December 25 was brief and indicated that one of the crew members of the *Mary Sachs* was missing. A search party of two sledges was sent out without locating him. Christmas was celebrated with a large meal in spite of the missing man. Chipman provided the details of the menu: "Olives, Endicott Soup, Roast Stuffed Duck, Giblet Gravy, Cranberry Sauce, Mashed Potatoes, String Beans, Christmas Pudding, Mystical Sauce, Mince Pie, Dates, Chocolates, Fruit, Coffee, Cigars, Cigarettes."[15] Leffingwell provided the duck and some whiskey. The evening was spent watching movies that had been taken by a CAE member.

The lost crew member was located the following day. He was extremely lucky, suffering only a bit of frostbite on his face and large toes. His hands and arms were not damaged but Leffingwell described his head as "muddled."

Leffingwell wrote very little about his month-long stay, but mentioned that he had helped with astronomical observations, played bridge, discussed politics (with a reference to Stefansson), and had been provided with material to repair the Flaxman house stove. He had actually accomplished more than he indicated in his journal. During his stay

Leffingwell also helped clean equipment, build a snow house, and place a roof for an outbuilding. He had many discussions with O'Neill, the CAE geologist, about the geology of the area. It was known at the time that the Brooks Range was an extension of the Rocky Mountains. Leffingwell shared with O'Neill that Alfred Brooks of the U.S. Geological Survey had proposed naming the Alaska extension "Arctic Mountains." Following the death of Brooks in 1924, it was named "Brooks Range" in 1925, the name used today.

Return to Flaxman

Leffingwell left Collinson Point for Flaxman Island on January 19, accompanied by Captain Bernard, topographer Chipman, and Cook, the crew member who had been missing at Christmas. The sledge trip was uneventful, covering the approximated 35 miles in seven hours. Leffingwell determined that the minimum temperature at the house had been -40°C. Bernard and the others returned to Collinson Point on January 22.

As remote as Flaxman Island appeared, it seemed that there were always people coming and going. On January 26, McConnell and an Inupiaq, Fred, came through on their way to Barrow and stayed the night. McConnell wrote: "sun returned (at Flaxman Island), on the 26th of January!...I have been able to see beacons at a distance of from six to nine miles, without the glasses. Leffingwell certainly has made it easy for travellers in this country."[16] The beacons that Leffingwell had established for his triangulation network aided those traveling the coast.

Leffingwell wrote letters home and sent them out with McConnell. Terigloo and his family arrived the following day. Leffingwell had arranged for them to accompany him on his field work, but they would also stay with him at the Flaxman house. Terigloo would hunt for food while Leffingwell worked on mapping.

According to Leffingwell's journals, Captain Bernard and others arrived from the west on January 31 with others and left on February 2 for Collinson Point. Leffingwell started planning his spring work and set his goals: "1st trip Feb 6–7 stations nearby. March-April 15–20 sta. to west as far as Kuparuk. April 15-30. Local & Kugruak [Canning] May 1–15 between here & Col Pt or farther if possible."[17]

Spring Field Work

The rigors of field work along the Arctic coast in the spring were vividly illustrated by Leffingwell's journal entries for February 14 through February 28, when he traveled to Challenge Island, Point Gordon, Kadleroshilik triangulation station, Heald Point, Stump triangulation station, and Tigvariak.

In preparation he made a new tent, sewed canvas for the tent covers, and bent sticks for the tent frame. He also worked on the camp stove, overhauled the theodolite, and sewed additional covers for the axe, fry pan, and other camp gear. Leffingwell iced the sledge runners and by February 14 was ready to go. He calculated the sledge weight at about 310 pounds, including about 50 pounds of food for 10 days. The following morning at 9:30 a.m. Leffingwell and Terigloo's 13-year-old son left the house, hauling the sledge and using the sail when the winds allowed; no dogs accompanied them. Terigloo's son primarily assisted with hunting and setting up camp and occasionally assisted with re-establishing signals for triangulation.

They stopped at several of his nearby triangulation stations to place new signals for those blown down, then camped at 3:00 p.m. Leffingwell wrote later that evening that the sledge weight was too much for one man and a boy. Most days the wind blew anywhere from 15 to over 45 miles per hour and on one occasion, snow fell for most of the day. Temperatures dipped to the negative 30s and they used the sledge cover as a blanket. Leffingwell spent the time inside the tent reading, playing solitaire, and drawing ships. On February 26 they moved to Aigukuk's igloo and had to dig out the door, as a snow drift had blocked the entrance. From Aigukuk's they returned to the Flaxman Island house the following day, leaving the theodolite and tripod at Aigukuk's. Two days later, Leffingwell was working on the stove and noted that the sun was out at noon and set at 4:52 p.m. He accomplished many tasks and wrote: "5 good days in house, after 11 bad ones in field."[18]

On March 5, he returned to Aigukuk's for the theodolite and tripod and continued on to Point Gordon for some cached flour of Terigloo's. They left on March 8 and camped in the afternoon at the Kadleroshilik rack triangulation site. They were pinned down in their tent by another blizzard and by temperatures in the negative 40s (the instrument froze at -42°). The water pot was frozen solid, Leffingwell awoke to frost on his pillow, and felt it was the coldest wind that he remembered ever experiencing. When he worked outside over the next few days, his fingers froze to his pencil and notebook. While experiencing bitter cold temperatures outside, Leffingwell was curious about temperature variation within the tent so he measured several distances from the stove and took readings, which varied from 1° to 17°.

On March 11, he observed an eclipse of the moon. Within a few days, gale force winds struck his camp again and Leffingwell judged their force in excess of 70 miles per hour. The winds scattered heavy boxes, including the telescope box, which he never found. He wrote: "Last night about 10–2am it blew harder I remember in tent ex [except] perhaps near Herschel [Island] in 1906. Ridge pole under outside cover sticks up & the calico flapped madly. Afraid it would tear. The flapping jarred bed clothes even. thot [*sic*] I might have to go out suddenly to fix up, so put on clothes & slept on bag with them. Only time I have done this in gale. Tent warm. T 8am = -22."[19]

After nine nights at the first camp they traveled 18 miles to the next camp near Heald Point on March 19. Over the following two weeks Leffingwell found that nine signals had fallen and had to be reconstructed. He took rounds of angles when the weather allowed, and read and played solitaire when it did not. In order to conserve fuel he used candles for light.

Leffingwell was camped on the west side of Prudhoe Bay when strong winds carried the top of the stove pipe off of the tent and deposited it about 250 yards away. On March 29 snow fell most of the day and he wore snow glasses as they moved to a new camp. They had two visitors the next day; a local trader, O'Connor, and a companion dropped by en route to Barrow from Demarcation Point. They spent the night and departed early the following morning. It was now April and the work was slow and tedious.

Leffingwell finished reading his last book and resorted to playing solitaire for diversion. The food supply had diminished to three days, assuming that they secured some ptarmigan, so the following morning, both Terigloo and Leffingwell went out hunting but did not record the results.

At the Tigvariak camping site, they stayed two days and Leffingwell was pleased to see a signal that he had placed in the spring of 1912 still standing. He also calculated angles and was gratified by the results.

Flaxman Base Camp

Leffingwell arrived back at the Flaxman Island house on the afternoon of April 21, after traveling about seven and one-half miles, much of it in deep snow and stopping along the way at two more triangulation stations. He had been away for seven weeks and described himself as "very thin." "My hands & feet have been cold in a way I never experienced before. Either I am loosing manly vigor with age or short rations & next to no grease in ration have cut down vitality. Hope latter... The Canadian outfit rides on sled with 10-11 dogs & I have been hauling like a horse. No more of it for me after this trip."[20]

Leffingwell was pleased to discover that McConnell of the CAE had left 20 pounds of butter for him. He polished off one-half pound immediately on 12 pancakes and ate again later that evening. Leffingwell settled back into the house with the usual chores. He washed and mended clothes, chopped wood, scrubbed the floor, melted ice, and "chopped ice out of corners." Even though he was exhausted, he polished off a couple cups of coffee, then had a difficult time sleeping, so gave up and started working at 3:30 a.m.

Overall, Leffingwell was pleased with the amount of work he had been able to accomplish in spite of the adverse weather conditions. He filled in gaps and noted a few areas where a bit more work was necessary. The new angles helped him uncover previous errors and improved the accuracy of his work. He took breaks from the calculations and chopped more wood, made a new box for the telescope, and packed up some boxes of gear and books.

Leffingwell was in his element working with the other scientists. He openly shared his knowledge and experience. They appreciated it and responded in kind. It was mutually beneficial. Chipman expressed his appreciation in a letter to a colleague at the Canadian Geological Survey:

> As I told you last fall I gave Mr. Leffingwell a note of introduction to you. It was purely on the personal basis letting those who know speak for his scientific work. If he ever presents it I wish you would take him out and give him one of the best dinners that can be found. I have

> enjoyed him as much as any of our arctic experiences and he has helped us out in our astronomical work, giving us short cuts and points learned only by experience which would have taken us a long time to learn for ourselves. Incidentally I have told him to feel very free to loosen up with you on the subject of V.S. [Vilhjalmur Stefansson] so if you get about three drinks into him and ask a few direct questions you'll get interesting news.[21]

At about the same time, O'Neill wrote to his mother that Leffingwell had stayed with the CAE for a month at Collinson Point and that they had enjoyed his company. O'Neill helped Leffingwell with information on an area of the Sadlerochit River to which Leffingwell had not yet traveled. O'Neill had no intention of intruding on Leffingwell's efforts. And the CAE certainly had no intention of duplicating his efforts along the Alaskan coastline.

In close quarters it was also common to hear less appreciative opinions about others. Leffingwell shared his opinions as well. The Southern Party was not particularly fond of Stefansson. Leffingwell expressed his concern that Stefansson would ruin their opportunities to conduct their scientific work. There were competing agendas. The relationship between Leffingwell and Stefansson was one of contrasts. They respected each other, but had very different personalities; Leffingwell was introverted, and Stefansson was extroverted. They were both driven in their work, sometimes at the expense of others. Leffingwell was a perfectionist and was often rigid, while Stefansson appeared more casual. Leffingwell did not seek attention, a sharp contrast to Stefansson. Out of their shared experiences in the Arctic, their respect for each other increased into their later years.

On April 29 Leffingwell and Terigloo's son were on their way to the Kugruak (Canning) triangulation station to finish the remaining work. Two days later, they camped three miles north of their destination. The snow was still drifting and plentiful, but Leffingwell finished taking angles and placed a new signal, to replace the old one that had blown away.

They returned to Flaxman the evening of May 3. Travel was easier than he expected, the load was lighter, and the weather was warm enough that Leffingwell wore a sleeveless undershirt part of the distance. He later wrote: "Pretty tired but not having to haul much, not so bad as at any camp going inland. Sun. Big change inland. Many bare spots, none going up. Boots soaked, but do not need them any more I hope. Soles gave out, of last pair."[22]

The next day Leffingwell recorded a temperature of 4.5°C. Everything was thawing quickly and he cleared the snow off the house and shed so it would not leak through as it melted. He switched his apparel to "water boots" and warm weather clothes. The next several days were beautiful and he took more rounds of angles, more photographs, shoveled more snow, deepened a trench for runoff, and packed some boxes of books. He mended socks, greased his boots, worked on calculations, and spent eight hours using a sewing machine to repair and enlarge the tent cover. He then spent all of May 8 working on calculations of the data he had collected.

Leffingwell walked to Brownlow Point on May 10 to measure more angles for his triangulation network. He camped that night on the tundra above the bank, took more angles the following day from several locations, and was back at the camp by 6:00 p.m. He decided to leave most of his gear except the theodolite and walked back to his Flaxman base arriving about 11:00 p.m. Okalishuk and Toxiak arrived the next day from Collinson Point with some meat and a letter for Leffingwell from Anderson, head of the Southern Party of the CAE. That afternoon, Leffingwell, with their assistance, continued measuring and calculated angles from the triangulation stations near the house.

When Kissik arrived from Collinson Point on May 14, he reported that the Canning River mouth was flooded to the ocean. Leffingwell started planning to move his dory closer to the house in anticipation of melting ice and open water.

Collinson Point and the CAE

Aigukuk and Aiyaunak arrived from Collinson Point on May 17 and stayed for three nights. Leffingwell planned to return to Collinson Point with them and they departed on the morning of May 20, arriving at the point before 6:00 p.m.

Within a week, Leffingwell decided to leave for two weeks in the mountains. He was accompanied by Terigloo's son and another Inupiaq who planned to bring the sledge and dogs back. But the proposed trip was thwarted by rising waters from spring break-up in several places, which made it impossible to cross to the mainland. They returned to Collinson Point that evening and tried again on May 30, but still could not cross. This time they cached their gear and again returned to the point. The first few days of June were marked by snow, fog, and temperatures below freezing. Leffingwell and Terigloo's son both were sick with the flu, and for the next two days, Leffingwell recorded only the temperatures. Finally on June 9, they returned up the Katakturuk River to their cache, retrieved their gear, and established camp about a half mile away on the bank of the river. They returned to the cache for another load before calling it a day.

Leffingwell crossed the Katakturuk River the following morning. The weather was clear and beautiful and he wrote that he had found some petrified wood. By evening rain was falling in the mountains. The following day, they traveled seven miles before camping near a bluff where he placed a signal.

The river was still in flood on June 13 and Leffingwell's neck ached from carrying a heavy pack, so he decided to cache all the food and return to Collinson Point. He wrote: "15 ft of water in front of camp

Katakturuk River canyon, Sadlerochit Mountains. *Gil Mull Files*

& deep. Before it was dry channel. Level of water is about one foot below camp level, but is visibly going down."[23] They left some gear and the tent at the camp and about noon headed back to Collinson Point. Leffingwell had returned with "many insects (bumblebees, beetles, spiders, flies, etc.)" for Johansen, the CAE zoologist. The following day, Johansen traveled with Leffingwell to the east and helped him collect fossils. They explored the Carter Creek area and collected some fossils. On June 20, Leffingwell spent the day visiting with Anderson and two other members of the CAE. Leffingwell spent a couple of additional days reading and measuring more angles for his triangulation network.

On June 24, a party of ten that included Leffingwell and Anderson departed Collinson Point for various destinations. Leffingwell, Nehmens, and Louis planned to travel to Barrow from Flaxman for their departure to the lower 48. Anderson and two others planned to travel only to Flaxman to retrieve some gear, while the remaining four were headed to Konganevik Point for hunting. They left at 11:00 p.m. in the mid-summer nighttime daylight and established camp southwest of Konganevik Point at the southern end of the bay at 6:00 a.m. Along the way, Leffingwell stopped at the Konganevik triangulation location to measure more angles.

The next evening, Leffingwell mapped the coastline about one mile east of camp before they continued west. They traveled about 17 miles to the Brownlow Point triangulation station, where they made camp. During the night and early morning, he finished mapping part of the spit not completed in 1912. He finally turned in about 1:00 p.m. and slept for twelve hours straight; up and moving again at 1:00 a.m.

Return to Flaxman Island

On June 26, Leffingwell, Anderson, and the others traveled the remaining five miles to his Flaxman base camp, where they arrived about 9:30 a.m. There was still solid ice south and west of the house on July 1. Nehmens and Louis of the CAE were at Flaxman to accompany Leffingwell aboard the dory to Barrow. Their travel would require rowing, sailing, poling, and taking soundings. As they waited for the ice to move out, Nehmens made a rudder and a mast for the dory.

Leffingwell was still trading with the Inupiat; taking caribou skins in exchange for items he would not take out with him. He spent the first ten days of July washing, cleaning, packing, and reading. Leffingwell measured a few more angles and took additional photographs of the ground ice on the northern side of the island. He monitored the movement of the pack ice or lack thereof; he found that when the winds blew from the west or southwest the ice moved out from the shore, but when the winds shifted, the ice returned to shore. Occasional snow and fog delayed their departure. On July 10, there was too much wind for last minute mapping so he took photos of the ice on the north side of the island.

Leffingwell had been meticulous in his literature review and examination of ground ice, commonly referred to as permafrost—permanently frozen ground. He focused on coastal areas. He wrote that it was only in his last year in northeastern Alaska that he came to understand the formation, structure, and distribution that challenged previous theories. He wrote that "poor exposures" had made study difficult. In previous years, he had seen a partial perspective of ground ice formation, but not the entire picture.

Leffingwell found that ground ice occurred in unconsolidated sediments, not in hard rock. He found that it occurred more often in clay and silt or muck than in areas of sand and gravel. He also determined that there were many types of ice and an occurrence might be composed of one or more types. And he later discussed in his report the processes that would account for ice preservation; through gravity, such as slumping, or by material transported by water in spring floods.

Traditional theory suggested that snow or ice had been buried by and under other material, and thus preserved, occurred in continuous sheets. But when he observed a vertical wedge of ice, he found that it had actually formed underground. He found in northeastern Alaska that ground ice occurred in a "network of vertical wedges." Leffingwell also determined that its upper edges were close to the depth of thaw.

He described frost cracks found in silt and sand and wrote that the cracking ground was heard "everywhere." He noted that once the snow melted, the cracks in the ground developed polygonal patterns, appearing as pentagonal, hexagonal, and rectangular blocks. He estimated that the blocks he observed had an average diameter of 16 yards.

He summarized his findings by detailing the six types of ground ice he had observed: grains mixed with earth, thin sheets amidst earth, clear

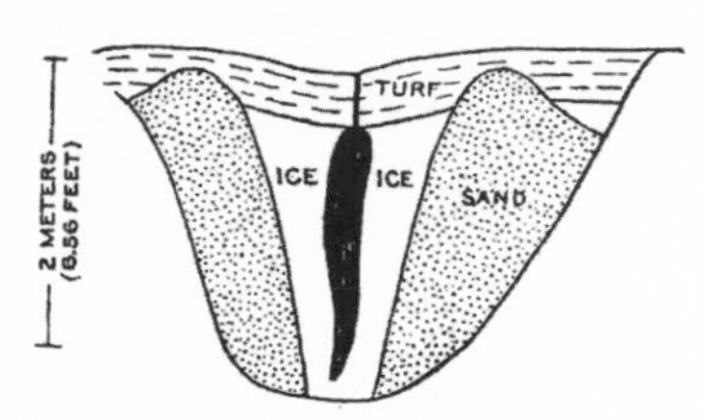

FIGURE 21.—Ice wedge in sand. A tunnel has been cut in the ice by drainage of surface water through the frost crack. The sand on either side is apparently bulged up.

Ice wedge in sand

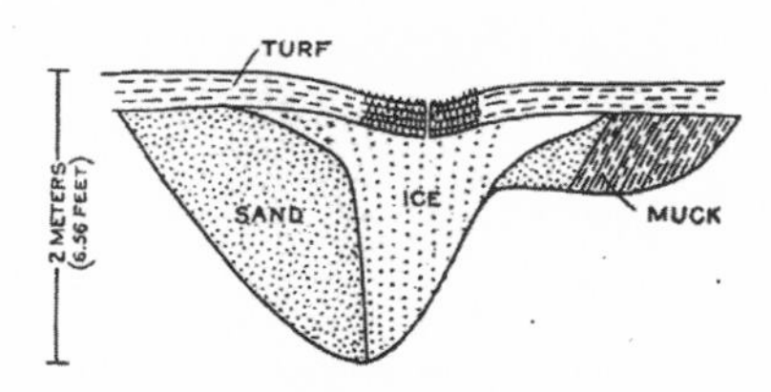

FIGURE 24.—Large ice wedge which spreads out under the surface of the ground. The vertical lines indicate rows of whiter ice full of air bubbles. The material on both sides is sand. To the right are upturned muck beds.

Large ice wedge

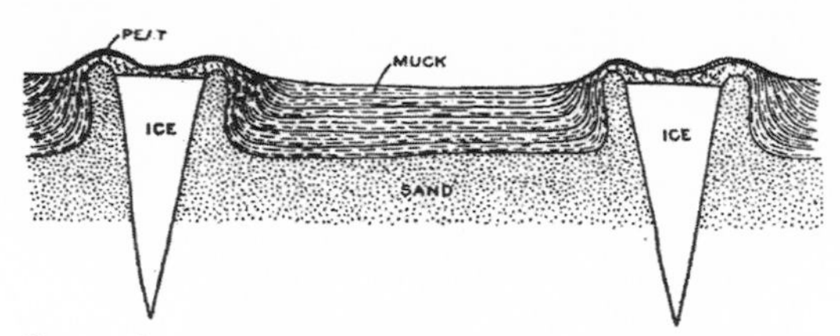

FIGURE 25.—Hypothetical section of ice wedges and depressed polygonal block.

Hypothetical section of ice wedges

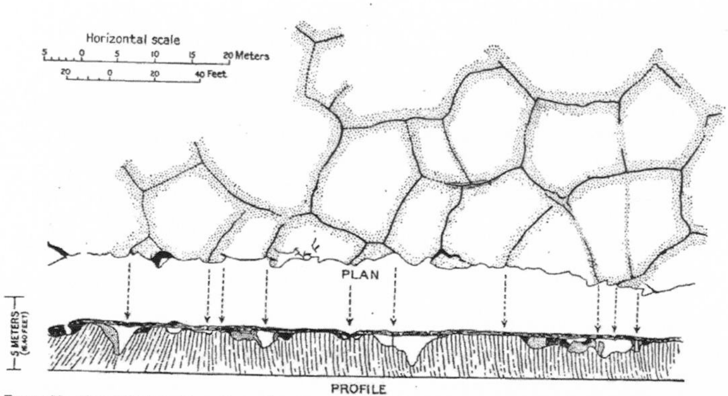

FIGURE 26.—Map of frost cracks on the tundra, with a sketch of the exposures of ground ice in the bank at one edge of the mapped area. Heavy lines represent frost cracks open in July. The dotted lines show evident loci of frost cracks. Stipple marks show areas probably underlain by ground ice. In the section below the map white areas represent ground ice; dotted areas, sand; lined areas, upturned muck beds. The rest of the exposure has slumped.

Frost cracks on the tundra

Leffingwell sketches of ground ice (permafrost) and ice wedges. *USGS Professional Paper 109*.

horizontal ice beds, beds of ice amidst beds of earth, ice deposits with some isolated earth materials, and vertical ice wedges.

Leaving Forever

Leffingwell had returned to the North Slope in 1913 to retrieve instruments and finalize mapping of the coastline. With that accomplished, he made final preparations to leave Flaxman. The trip to Barrow would be full of lengthy delays, trials, and tribulations, marked by adverse weather, incidents involving their boat, lack of sleep, food shortage, and the real possibility that they might miss their transportation to the lower 48 and be forced to spend another winter.

On the morning of July 11 they were finally able to leave for Barrow. Leffingwell wrote: "Up 6:00am. clear warm & 10-12 ENE wind. Off 9:00am. Closed up house, boarded windows etc. Left a few cases freight for Terigloo or *Anna Olga* to bring to Barrow. Couldn't take all in dory."[24] Leffingwell also left behind his personal library at the Flaxman Island base camp. It was an extensive collection and Stefansson later wrote: "The shelves were full of scientific books. I read Osler's 'Practice of Medicine,' in its marvelous first edition and Chamberlain [*sic*] and Salisbury's three-volume geology. Then there were books labeled romances, such as the marvel stories of H.G. Wells. There were whole shelves of Tolstoy and of the English classics."[25]

Whatever thoughts Leffingwell had about leaving the house at Flaxman Island and the North Slope for the last time, he kept to himself. They traveled about three miles to Mary Sachs Island (which no longer exists) and camped at noon. They were then forced by ice conditions to stay for nine nights before they could continue. Leffingwell spent part of the first afternoon mapping the south side of the island. He carefully monitored ice conditions and they were finally able to leave on the morning of July 20. They made some progress then spent a week waiting for travel conditions to improve and the ice to clear out. Leffingwell spent some of his free time mapping the polygons on the tundra and drew sketches of ice exposure in a bank near camp. He also spent a lot of time in the tent. He was tired of all the delay but was forced to wait for ice and travel conditions to improve.

By August 5 they were en route to Harrison Bay and Cape Halkett, northwest of Prudhoe Bay. They camped on the tundra near Terigloo for much needed sleep. Leffingwell had been up for 54 hours straight and wrote that he had worked for 50 of those hours. Half of the time was spent sailing and he spent 15 hours at the helm. Sailing required careful attention as they encountered small pieces of cake ice during their journey.

Leffingwell slept for nearly 20 hours and wrote that he "could have slept longer." The next day they were able to use the sail at times and by the following afternoon they caught up with Terigloo (who had the remainder of his gear). The party reached Pitt Point on August 8 after traveling about 15 miles in eight hours.

There was too much ice on the outside of the islands for Leffingwell's comfort. They observed heavy ice to the west and slowly poled their way near shore until they decided to camp. As the party continued west, Leffingwell took soundings when he thought it necessary and they set sail when possible. Occasionally he observed whaling ships or canoes passing to the outside of the islands via the leads in the pack ice. He was concerned about safety, and did not want to spend 24 hours a day in the

Coastal plain polygons, 1992. *Janet Collins Collection*

boat traveling on the north side of the islands with no opportunity to land.

At one camp, waves hit the dory in the middle of the night, so Leffingwell pulled the boat up further ashore and secured it once again. When he awoke the next morning, he discovered that waves had washed around their tent but not in because they had banked up sand around the tent. The boat was full of water. Everything was wet: instruments, records, letters, and film. Leffingwell pulled some of the items out for drying, but decided that the remainder would have to wait until later.

The food supply now consisted of pancakes, a small amount of bacon, and tea. They decided to go hunting and when they returned to camp, the boat was "high & dry." The party had to unload the boat to move it and Leffingwell estimated that the load weighed about three tons. Then it snowed. They departed at 8:00 a.m. into swells, currents, and winds. They camped that evening on the southwest end of Cooper Island, after 13 hours and 25 miles. (Leffingwell referred to it as Iglura Island in his journal.)

The party set off by 1:00 p.m. the following afternoon and before supper reached the *Jeanette* anchored near the Tapkaluk Islands east of Point Barrow. Because of the ice, the *Jeanette* had been anchored for 10 days and unable to reach Brower's at Barrow. The winds shifted that evening and Leffingwell, Nehmens, and Louis rowed to Point Barrow. They finally reached Charlie Brower's place about midnight and walked to Hopson's for a meal and some sleep. That night the winds blew from the west and southwest, which carried the ice away from the shoreline. Brower sent dories, canoes, and whale boats out to the *Jeanette* about a mile east of Point Barrow and towed her in to shore.

Leffingwell's final journal entry was for Wednesday, August 19, through Saturday, August 22. He wrote about ice conditions and whaling ships arriving at Barrow. He took passage on the *Jeanette* and the ship departed at midnight on August 22.

It must have been a bittersweet departure. For nine summers and six winters the Arctic of northeastern Alaska had been his home. His life had been greatly enriched by his many experiences and the Inupiaq culture. His love for the Arctic landscape never diminished.

It was now time for Leffingwell to move on with his life. He had finished the mapping of the coastline that he had set out to do and recovered his instruments. Yet he was left feeling that he had not accomplished very much for his time and money. Others would contradict those conclusions. In fact, he had been part of the Baldwin-Ziegler Expedition attempt to reach the North Pole, co-commanded the Anglo-American Polar Expedition, determined with Ejnar Mikkelson the edge of the continental shelf, accurately mapped the coastline from Barrow to the Yukon border, accomplished pioneering work on ground ice, defined and mapped the geography and geology of a significant portion of what is now the Arctic National Wildlife Refuge, provided nomenclature for much of that area and attempted to honor the local Inupiaq names, named the Sadlerochit Formation reservoir underlying the Prudhoe Bay Oil Field, written about equipment, food, shelter, and clothing, and recorded the area through his photography and illustrations. His efforts required travel of 4,500 miles over 30 months, on foot, or by boat or sled, and he had established about 380 camps. Leffingwell calculated that his three trips on the ocean totaled 20,000 miles—this from a man who suffered from seasickness. And now, it remained to be seen what his next journey in life would be.

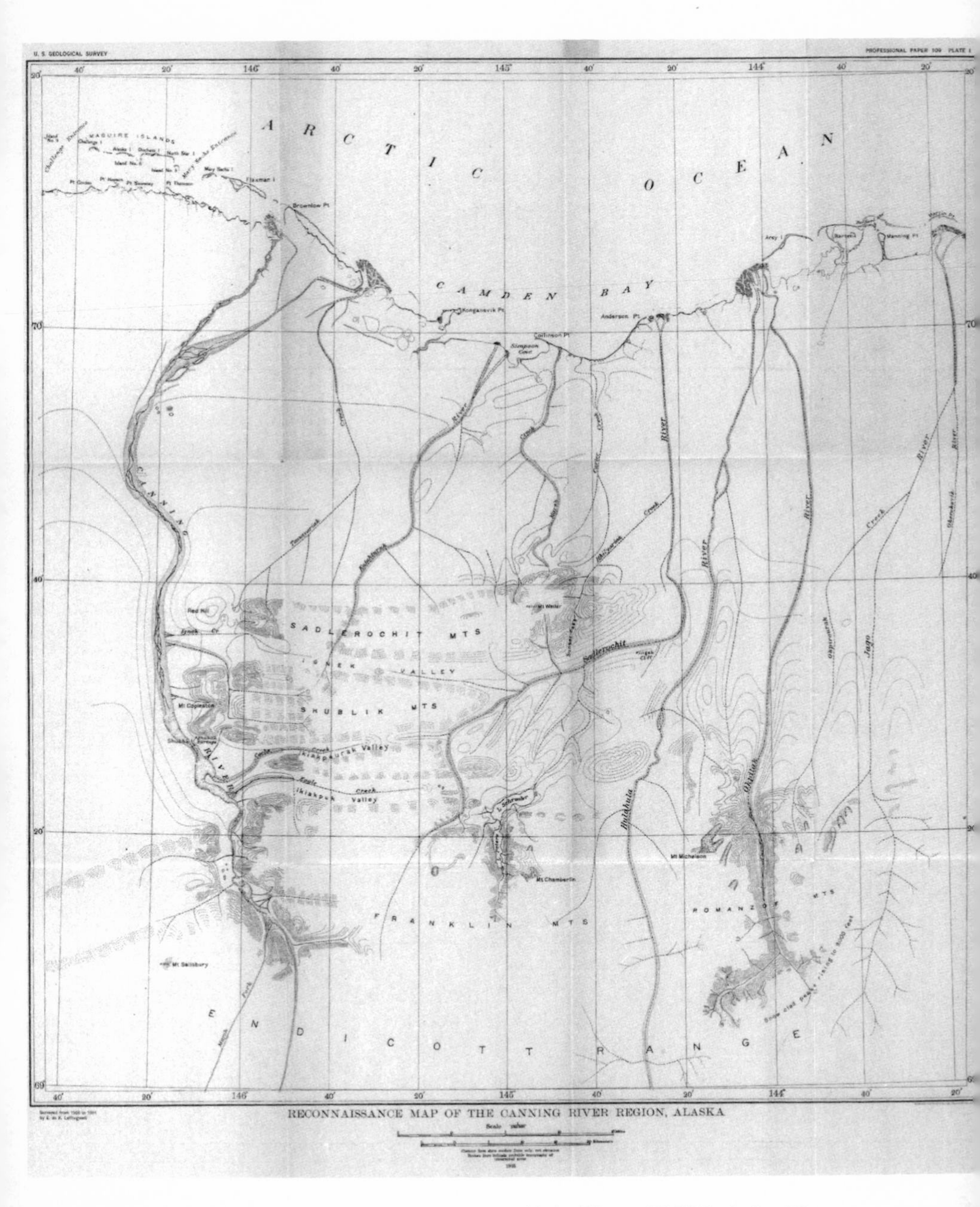

Reconnaissance map of the Canning River Region, Alaska, Plate 1. *USGS Professional Paper 109*

CHAPTER 12

Post-Arctic Life: 1914-1949

Leffingwell left the Flaxman Island base camp and the Arctic for the last time on July 11, 1914. Upon his return to the lower 48, Leffingwell walked into the office of Alfred Brooks at the U.S. Geological Survey in Washington, DC, in 1914 and told him of his nine summers and six winters mapping in northeastern Alaska. Brooks gave Leffingwell an office where he spent 18 months writing up the results of his research. The outcome was Professional Paper 109, titled *The Canning River Region, Northern Alaska*, a detailed report on the geography, geology, ground ice, meteorology, hydrography, nomenclature, and mapping of the region. Professional Paper 109 represents Leffingwell's largest achievement and focuses on the northwestern section of the current Arctic National Wildlife Refuge and the coastline area extending from Barrow to the Yukon border. The Professional Paper also contains particular detail on scientific equipment, clothing, food, shelter, and equipment, including sledges and dog harnesses. Since its publication, the paper has been a particularly helpful reference for scientists working on the North Slope of Alaska.

When Professional Paper 109 was finally published in 1919, the content created a stir, in particular, the section on potential petroleum reserves. Oil companies rushed to the North Slope to stake their claims. President Harding signed Executive Order 3797A on February 27, 1923, setting aside the area west of Prudhoe Bay as Naval Petroleum Reserve Number 4. Now known as the National Petroleum Reserve Alaska (NRPA), the reserve was established for oil and gas reserves only, and currently encompasses 35,984 square miles.

The Executive Order acknowledged: "large seepages of petroleum along the Arctic Coast of Alaska and conditions favorable to the occurrence of valuable petroleum fields on the Arctic Coast and whereas the present laws designed to promote development seem imperfectly applicable in the region because of its distance, difficulties, and large

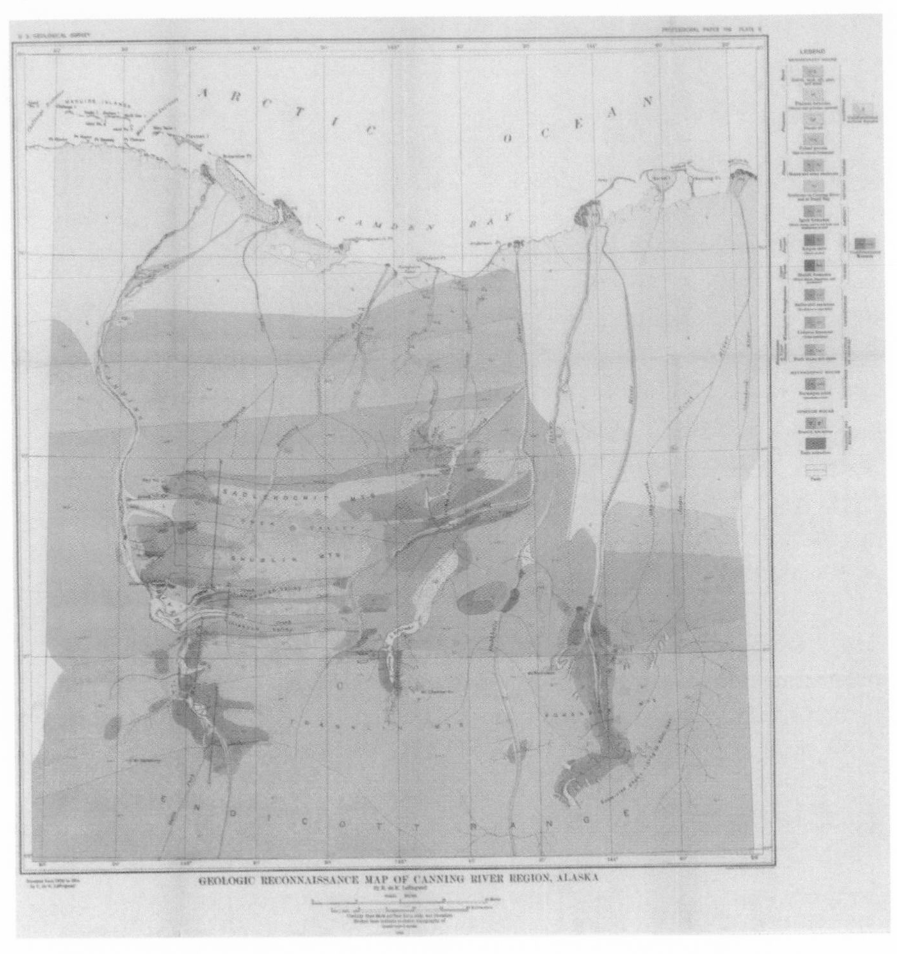

Leffingwell's geologic reconnaissance map of the Canning River Region, Alaska, Plate 2. Note the correlations in rock types, banding, and physiography with the more recent map on the facing page. *USGS Professional Paper 109*

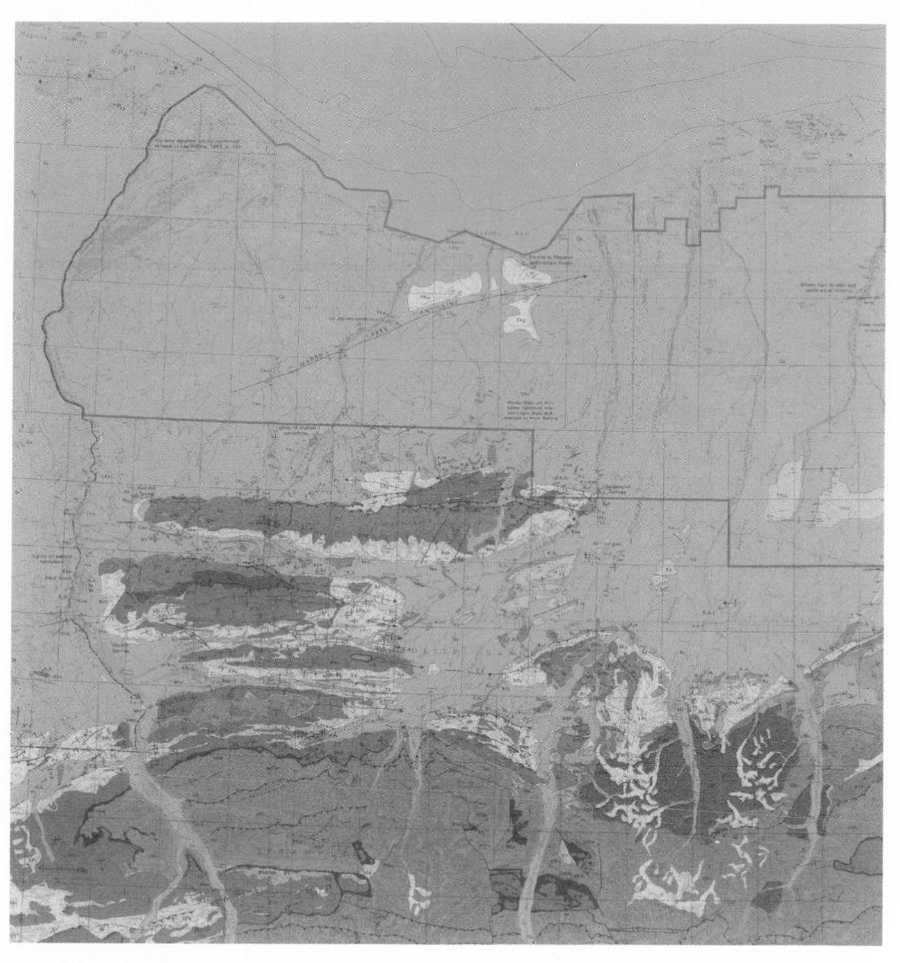

Current geologic map of the Demarcation Point, Mt. Michelson, Flaxman Island, and Barter Island Quadrangles, Northeastern Alaska, 1986. *USGS Geologic Investigations Series, I-1791.*

expense of development and whereas the future supply of oil for the Navy is at all times a matter of national concern."[1]

Professional Paper 109 was not the first to report the oil seepage near Cape Simpson, southeast of Barrow. Leffingwell had collected samples and communicated with Alfred Brooks of the U.S. Geological Survey about "petroleum residue" located near Smith Bay, about 100 miles east of Point Barrow. Inupiat were well aware of it, as were some of the whalers. Brooks wrote about it in the 1908 publication, *Mineral Resources of Alaska*: "These fragmentary data point to the conclusion that there may be a petroleum field in this extreme northern part of Alaska. Were the region not so inaccessible, it would certainly be worth while to investigate these occurrences, but as it is, even if petroleum is found, it could not now be brought to a market."[2]

Until his return to the lower 48 in 1914, Leffingwell had focused his life and work in northeastern Alaska; that would now change forever.

Anna May Meany

During the time he was working on the professional paper, he met and fell in love with Anna May Meany. They married in 1916. They met through family friends, corresponded regularly, and visited occasionally while he was working on the Professional Paper. She wrote about it to Gil Mull in 1973: "I first met Ernest Leffingwell in 1915 in Washington DC during my Easter vacation while visiting my aunt and uncle Mr. and Mrs. John Storrow. At the time I was teaching school in New York City...My family had met Ernest some years before, during one of his earlier visits there and had heard of their Arctic Explorer neighbor, who slept [outside] under a single blanket in order to toughen himself for the intense cold of Alaska."[3]

Leffingwell wrote to Anna at least every couple of days and, fortunately, she kept his letters. They provide insight into Leffingwell's character, perceptions, dreams, and goals. Leffingwell was 40 years old when he wrote the letters to Anna, and often signed his letters as P. Bear or Polar Bear or Polar Bear of the Den.[4]

Leffingwell's journals from the expeditions and years on the North Slope of Alaska rarely showed emotion. They tended to be direct, clinical, and a true accounting of his work and how he spent his time.

But with Anna, his writing was quite different. Leffingwell was clearly smitten and distracted from his work.

In one of his early letters to Anna he described her in geologic terms:

> Who cares about the thickness and age of some pesky old rocks, when there are such interesting speculations as to how old is—somebody nice. I am afraid that I shall be saying that the Carboniferous formation is about five feet four inches tall, the color fair, texture, very soft, the age, just right, structure, lovely; occasionally volcanic, but usually joyous, some granite in the composition, but also very much pure gold; heart ranging from marble on the surface to molten lava at great depths, which the investigation has at present been unable to sound.[5]

He was very much a romantic, continuing: "I shall throw a kiss to the moon every time I see it. If you see the man wink at you, it means that he has just received one from me and would like one from you to keep it company."

Leffingwell's letters also offered insight into his life and his sense of humor. He was not certain that Anna would respond positively to the seriousness of some of his letters. "It will be after you have received some of these too serious letters of mine, and I shall be uneasy until I know whether you are going to tell me to go and lecture my grandmother about how to suck eggs. But you won't take it amiss, if I know you."

Prior to meeting Anna, Leffingwell was uncertain about his future once he finished his report. He was feeling unsettled and thought of re-enlisting in the military. But as his feelings for Anna grew, Leffingwell wrote to her that he no longer felt that way. They worked on their relationship and the uncertainties that accompanied it.

Leffingwell was focused on finishing his report, but was also concerned about not having "visible means of support." Financial support for his work in Alaska had been provided by his parents. Leffingwell was ready to move on with his life and he understood that he would be starting anew. He felt that he had not kept up with developments in geology. And Leffingwell felt that it was important to remain humble about his experiences in Arctic exploration. What concerned him the most was that he was unsure how Anna would respond to not having the abundant financial resources to which she was accustomed.

Leffingwell struggled at times expressing himself to Anna. "It is easy work to describe rocks and rivers, but emotions require an abler pen than mine." He worried about their age difference and her mother's cool reaction toward him. Leffingwell had spent many years alone and lonely in Alaska. He was 15 years older than Anna, relatively inexperienced with women, and lacked confidence in his interactions with people. He confided to Anna that he felt uncomfortable in social situations. Leffingwell was surprised by how easy conversation was with Anna, even from their first meeting.

He rarely interacted with anyone who understood what it was like to be in the Arctic. Leffingwell's independence and relative isolation in the Arctic may easier explain how he described himself in letters to Anna. He doubted that people liked him, but if he sensed that they did like him, he felt like he was too friendly. He also admitted that he could be cold and angry if verbally attacked. He was particularly introspective. In his letters one senses that he wanted to explain all aspects of his personality to Anna so she would know him as he saw himself.

In a letter written in September, Leffingwell was also realistic about the challenges of marriage, noting that "happy marriages are rare." He acknowledged that both of them were strong willed but felt if they were not selfish, their relationship would work. In a rare moment, Leffingwell described the challenges and occasional conflicts that arose as joint commander with Mikkelsen on the AAPE. "Mikkelsen sometimes swore at me in front of the crew, yet I held my peace and next day quietly took him to task, and always succeeded in making my point."

Leffingwell could be controlling and frustrating, no doubt. He had been raised in a very structured and rigid environment with strong beliefs about morality and religion, and they defined his values. It was clear to him how people should live their lives and he tended toward judging others, but not always openly. His interest in science and precision in all respects extended to personal habits. One example was detailed in a letter to Anna: "I have been keeping track of the time it takes me to bathe, shave and dress and this morning I was ready for breakfast in twenty minuits [minutes]. This is too fast for comfort, and I shall set thirty minuits as a standard."

But his love for Anna took precedence over everything else. He wrote another article for the University of Chicago alumni magazine in which he indicated that he was working for the U.S. Geological Survey and hoped to complete the work by summer. Leffingwell was quoted: "I do not think that I shall go north again, as I have very little to show for the last eight or nine years which have been entirely devoted to this work." He underlined the sentence in pencil, "I do not think that I shall go north again" and noted at the bottom of the copy, "I have also another very good reason which it will not do to tell. This reason has received a tremendous significance since 8pm Monday, April 5, 1915." It was undoubtedly a reference to his meeting Anna and their future together. Disconcerting was his erroneous attitude that he had not accomplished much work during his time in the Alaskan Arctic.

Anna encouraged and validated him in his efforts. Leffingwell greatly appreciated her kind words, but was clear about his attitude toward self-promotion. He opposed the idea of someone promoting himself for fame or fortune. Leffingwell maintained a low profile and felt that those who knew and understood his work would appreciate his efforts. In later years, he would occasionally express disappointment that his work was not better known or appreciated.

USGS Professional Paper 109 in Progress

Anna was the motivator behind Leffingwell writing and completing his report. He realized that the sooner he finished it, the sooner he could turn his total attention to her. Leffingwell confessed that he worked well under pressure but loafed when time was not an issue and described himself as "intellectually lazy."

On a Thursday evening in August he wrote to Anna with specifics about the increasing pace and progress of writing his report. In another letter he wrote that he was tired of geology and would rather study her. He had finally finished coloring the geologic map, and in the process, had used 14 different colors to represent the different geologic units (see page 226). Leffingwell moved on to other tasks including coastal maps which he indicated were easier. In August 1915 he wrote: "Yesterday I copied out my results and brought my logarithms home, as all the calculating is finished. I have the results of six years of observations on three sheets of paper. They are the Latitudes and Longitudes of about fifty places on my map. I shall send them in to be typed and keep a copy here in my room for safety."

Once Leffingwell finished the maps he had started in April, he tackled the section on history and exploration of the region which involved more maps and text. He wanted to make certain that the original names were represented. Leffingwell made three trips to the Library of Congress to look up maps, place names, and information that was not then available at the U.S. Geological Survey. He calculated that it would take about a week to add new names to the maps and there would be about 200 place names represented. In his methodical and meticulous approach, he traced the origination of all the place names he wanted on the map and in the Professional Paper. Leffingwell also wrote Anna that the place names would need to be approved by the Federal Board on Geographical Names, a practice that continues today.

For the text portion, he worked his way through ten volumes of exploration history, although he only identified one of them. It was part of the British Blue Books, otherwise known as the Parliamentary Papers for the House of Commons and House of Lords. Leffingwell enjoyed reading expedition accounts and learning about the history of the place names previous expeditions had mapped.

Once he focused on writing, Leffingwell wrote between 12 and 18 pages per day. He had close to 700 pages, but felt that 500 needed rewriting and half of those with minor corrections. Leffingwell estimated it would total about 800 pages and take fewer than two months to complete. When Professional Paper 109 was published in 1919, it totaled 251 pages plus maps. Leffingwell had researched and mapped an area he calculated at 70 square miles. It was a comprehensive accounting of what Leffingwell had learned and accomplished in the Arctic. The majority of the paper concerns his research of the geography and geology. But it was not just defining the geography and geology of that area. Leffingwell's Professional Paper 109 included food, clothing, shelter, equipment, mapping techniques, climate, flora and fauna, population, transportation, history, and nomenclature.

Leffingwell estimated costs at $30,000 over the nine summers and six winters, although he wrote that $15,000 was expended during the first year, 1906-1907. His parents had provided the funds. Leffingwell attempted to offset some of the expenses by trading and whaling, but earned only a small amount by the former and nothing at whaling. As he contemplated marriage, his financial resources were minimal and that needed to change. He decided to work on the family ranch in California. Leffingwell had received offers of employment from the U.S. Geological Survey, but declined the offers. The low salaries and his concern about not being current with developments in the discipline of geology were the rationale he offered.

Wedding Plans

Prior to their wedding announcement, Leffingwell's father corresponded with Anna and welcomed her to the family. Anna's parents were wealthy and Leffingwell's father expressed concern that Leffingwell's simple style might not be compatible with how Anna had been raised. He also offered unsolicited advice to her about his son: "his lonesome life in the North, where he had no one to consider but himself, may have made him seem less sympathetic than he really is. Be patient with him, if he falls short of your ideal, & be perfectly frank with each other."

As supportive as Leffingwell's father had always been of his son, surprisingly, he wrote to Anna on November 25, 1915, regretting that he would not be able to attend their wedding, but would be in Pasadena about December 10 to "locate and begin the cottage." The record of signed witnesses to the ceremony does not include Leffingwell family members. It is not known if Leffingwell's family members supported the marriage. It is also not known if the cottage was built, as other dwellings are referred to in later correspondence.

Leffingwell may have been 40 years old but his father was clearly still offering support and guidance. "I had a very nice and encouraging letter from my father today in which he says that he thought I might get upon my feet in a couple of months...The whole tone of his letter is so fine that it makes me very happy and assured that things will shape themselves beautifully for us." Leffingwell felt an obligation to repay his father for the generous financial support. Nothing further would be given to him. Leffingwell continually expressed to Anna that things would be easier after he proved himself. He needed to work hard and earn respect, which he did without hesitation.

1916-1919

When announcement was made of his marriage plans, newspapers indicated that Leffingwell was a "sought after guest" in Pasadena society.

About the same time Leffingwell was quoted in the Whittier, California, newspaper: "Insofar as I am able to forecast my future...I am done with exploring. I had reached that point in life where Ossler [*sic*] said a man should be chloroformed.[6] Therefore it seemed advisable that I determine whether or not I would continue in scientific work or seek some other vocation. I considered seriously taking up scientific work for the government, but after mature deliberation concluded that the pay was insufficient to maintain the standard of living such environment calls for. My marriage was the deciding factor."[7]

Leffingwell and Anna were married at St. Luke's Episcopal Church in Evanston, Illinois, near Chicago, on January 12, 1916. Leffingwell traveled west to Pasadena, while Anna remained in the Midwest. He was one day shy of turning 41 and Anna was 25 years of age. By the Christmas holidays of 1916, Leffingwell and Anna were together in southern California.

During that winter, on a weekend outing in the mountains, they were caught in blizzard conditions on the upper reaches of Mount Wilson in California's San Gabriel Mountains. As they descended, they managed to cross the swollen stream and at times, Leffingwell carried Anna on his back. They were out for seven hours before reaching Pasadena in the afternoon. The outing was written up in newspapers and even his colleagues at the University of Chicago learned of the incident.

He and Anna lived on his father's citrus and nut ranch in Whittier until 1930. Charles W. Leffingwell had originally purchased 500 acres in 1888. The ranch was planted with lemons, oranges, and walnuts. The Leffingwell Ranch eventually consisted of 640 acres and was bounded on one side by what is today known as Leffingwell Road. After his final return from the Arctic, Leffingwell worked the ranch alongside his father and brother, Charles.[8]

In June of 1973, Anna wrote to Gil Mull:

> I would say, my husband probably should never have left the scientific field for business but we wanted to get married. Saleries [*sic*] were so small in those days, so we accepted his father's offer to live on the Leffingwell Ranch near Whittier and learn agriculture. Ernest had spent all his own money including most of his patrimony from his father's future estate on his work in Alaska. Ernest could have staid [*sic*] on at the Geological Survey in Wash. D.C. or have accepted an

> offer from Cal Tech [Throop College of Technology] in Pasadena [as] the chair of geology—both at small salaries. Also he felt he was not exactly up to date on Geology after all of those years in the north.[9]

Leffingwell's attention shifted to marriage, family, and agriculture, and away from the scientific world he had known. He had been a founding member of The Explorers Club and a member of other geographical and exploration organizations. But his priorities had changed and in a letter to The Explorer's Club dated March 10, 1916, Leffingwell resigned as a member of the Board of Directors. In their regular meeting minutes, his resignation was read "and accepted with regret." No further explanation was noted.

Anna Meany Leffingwell and daughter Nancy in 1918. *Deborah Storre Collection*

In December 1917 Anna and Ernest's daughter Ann [Nancy] was born.

The 1920s

As Leffingwell and his family settled into life at the ranch, he began to receive national and international recognition for his work in the Arctic (discussed in Chapter 14).

His mother, Elizabeth Francis, passed away in November 1926, survived by husband, Charles W., two sons, Charles Warring and Ernest deKoven, and two daughters, Hortense Nesbit Wilson and Gertrude Vaughan.

In the spring of 1928 Leffingwell and Anna joined two other couples for a cruise yachting in the Caribbean for five weeks. The *Sciala* was 225 feet in length and had a crew of 34, and had been owned at one time by Henry Ford. They started from Miami after arriving from Los Angeles via train. Their days were spent taking in the beauty of the Caribbean, swimming, being hosted by prominent local residents, and dining on plenty of good food and wine. Some days they abandoned their rooms for cool breezes and to sleep on deck.

Later that year, in 1928, Leffingwell's father passed away at 87 years of age. Leffingwell was 53 years old. His father's death must have had a profound effect on him. During his years in the Arctic, his father had always been especially proud of his son and his endeavors. He had been untiring in writing letters in support of Leffingwell's work in the Arctic, making sure the younger Leffingwell was recognized for his efforts, largely through the media.

Just prior to the end of the decade, Leffingwell and Anna adopted two very young children, a brother and sister named Eric and Christine.

The 1930s and 1940s

When the stock market crashed in 1929 and the 1930 depression hit, Leffingwell sold his interest in the Leffingwell ranch, and he and his family moved into a house in Carmel, California, owned by Anna's aunt and uncle, Anna and John Storrow. During 1931, Leffingwell was discharged from the Naval Reserve and received a pension of $24.00 a month. Finances were difficult in the 1930s and for some time in the late thirties they lived in Imperial Valley in southeastern California. Anna formally inherited the Carmel house in 1943. Leffingwell and Anna's marriage had its challenges and for a number of years they lived separately.

During World War II, the U.S. Weather Bureau (Service) contacted Leffingwell about his knowledge of Arctic climate. In June of 1943 they requested Leffingwell's meteorological records. He was one of the few people who had recorded detailed climatic data on the Arctic coast of Alaska.

As World War II continued, Leffingwell and Anna purchased two cars, a Ford Model B and Model A, which they ran on Blazo, or white gasoline, as regular gasoline was difficult to obtain. Leffingwell acquired property south of Carmel at Rocky Creek, and between 1941 and 1946, he built a house that still stands. He maintained a vegetable garden, fished in a nearby stream, and camped with his son, Eric. Eric became a Boy Scout and other scouts often joined them on camp outings. Eric's best friend, Clive Rayne, remembers using polar bear fur sleeping bags and how excessively warm they were. In 1946, the home and land were transferred to a private party.[10]

The following year, Leffingwell acquired another piece of property near the banks of the Klamath River four miles from the town of Klamath, California. Leffingwell was 72 years old when he began construction of another house. He wrote Stefansson in November of 1947, indicating that it was nearly completed. The location was a paradise for fishermen and Leffingwell spent his leisure time fishing. According to his eldest granddaughter, Deborah Storre, he also built a boat, a smoke house, and maintained a large flower garden of dahlias. Leffingwell and Anna traveled between Klamath River and Carmel spending extended lengths of time in both. While living in Carmel Leffingwell became interested in ham radio, using the call letters W6QFY.

In 1948 Ernest and Anna purchased a new Ford 4 door sedan. Leffingwell was 73 years old. They decided to "break the automobile in," and set off on a road trip to the East Coast that took six weeks and 7,406 miles, traveling through 23 states. In the pre-interstate highway system, travel was not quick or easy. Motels were $5 or $6 a night for two. They visited friends along the way, took in the countryside, experienced bad

Leffingwell and son Eric at the Klamath, California, house in the early 1950s. *Deborah Storre Collection*

weather in Denver, bought some chains for the car, and were glad they had them when they reached Donner Pass in California.

During the time that Clive Rayne knew them, his impressions of Anna were of her being very active in church and talkative, while Leffingwell was quiet. He described Leffingwell as meticulous and recalled that he had a "top flight" gun collection.

As the 1940s came to a close, Leffingwell was not quite 75 years old, about to embark on other journeys and adventures, and hardly slowing down.

CHAPTER 13

The Later Years: 1950-1971

In 1951, at age 76, Leffingwell was quite active and healthy. He even traveled to western Canada on a hunting, fishing, and camping trip. He was still in communication with Stefansson and Anderson of the Canadian Arctic Expedition.

In 1952, he had decided to reside in Carmel for six months of the year from October 1 to April 1, and in Klamath Glen for the remaining six months. Part of that time in Carmel was spent traveling. Leffingwell and Anna went to Europe for over four months in November 1952 via the SS *Dalerdyk*. As seasick as he always became, it was amazing he was willing to go. Rough seas troubled him well into the trip. The ship sailed from Oakland, California, and transited the Panama Canal. They arrived in Antwerp, Belgium, on December 15 to cold and snowy weather. They proceeded through Germany to Copenhagen, Denmark, arriving on December 16.

Leffingwell and Anna met his longtime friend Ejnar Mikkelsen for lunch on December 18 and they returned to the hotel for more conversation and to look at the maps Leffingwell had created for Professional Paper 109. Two days later, they "went to see Einar Mikkelsen and wife Ella in PM to Charlottelund about 1 hr out in suburbs. Had delicious dinner and a most pleasant visit." Anna continued in her diary, "Newspaper reporters and photographers came to interview Ernest about his work in Alaska. Home early with a hoarse throat after a very happy visit."[1] In the years following the Anglo-American Polar Expedition, Mikkelsen had a very distinguished career. He organized an expedition to northeastern Greenland; he was Inspector of East Greenland from 1933 to 1950; chairman of the Danish Arktisk Institut; a writer; and advocate for the people of Greenland. He also assisted the Americans in the location of airbases in Greenland during World War II.

From Denmark, the Leffingwells traveled to Norway, England, Ireland, and France before returning to the United States. They returned on the *Andrea Doria*, which they boarded in Cannes. The ship proceeded to Naples, then Gibraltar before heading across the Atlantic to New York City, arriving March 16. They traveled to Washington, DC, by train and then west to San Francisco, arriving home on March 24.

Anna Leffingwell, sometime in the early 1950s in Carmel, California. *Deborah Storre Collection*

Leffingwell turned his attention to the Klamath Glen property in northern California in mid-January of 1953, when a terrible flood roared down the Klamath River. The house was badly damaged, inside and out. When the waters receded, up to six inches of silt remained inside the home.

In early April 1954, Leffingwell and Anna returned to Europe. Leffingwell was 79 and Anna not quite 64. They again departed from northern California, transited the Panama Canal, and arrived in Antwerp, Belgium, on May 10. On arrival in Antwerp, they traveled to Dover, England, and finally arrived in London. Over the following month they explored southern England, including Turnbridge Wells where Leffingwell's mother had been born. Leffingwell also visited the Royal Geographical Society in London and the Royal School of Mines at Imperial College.

They crossed the English Channel once again and spent the next five weeks exploring Paris and the wine country of Loire. The SS *Scythia,* their ship to Quebec, departed from LeHavre, France, on July 23. From Quebec they traveled to Carmel via train, arriving home on August 4.

Back in Carmel, Leffingwell caught up on correspondence. No matter their differences in the Arctic, he and Stefansson had corresponded for years. Stefansson had written to Leffingwell in July 1955 encouraging him to write an autobiography. Leffingwell responded much later, without committing to the idea. He was not inclined to promote himself, but he did share that he had planted over 400 potted azaleas and tuberous begonias at the house in Carmel. They grew so well and provided so much color that tour buses in Carmel diverted by their house to view the impressive display.

A monstrous flood in late December 1955 carried away the Klamath Glen house completely. According to Leffingwell's eldest granddaughter, Deborah Storre, he had kept personal memorabilia and most of his trophies there, including those won by his springer spaniel, Nuthill Dignity, an international champion that had won 17 challenge certificates. All were lost.

In a 1957 interview, Leffingwell talked about building a "tunnel boat" with, minimal draft of about two to three inches. When he was in his early 80s, he finished rebuilding a car that he had purchased as a 1932 Model B Ford in 1950. The car used his Model B engine, but he cleverly used and modified other parts to finish the chassis. It was a station wagon with wood trim and painted black. In a letter to Stefansson, Leffingwell made reference to a news article that referred to him as "Hot Rod Leffingwell, the Octogenarian Speed Demon." The article ended with this quote: "My father lived to be 87. I used to think I would too. Now I'm going to live to 102 if I die in the attempt. Once I quit being active I'm sunk."[2]

In May 1958, Leffingwell wrote Stefansson again in response to questions about ice islands and ground ice. His mind was still sharp and he enjoyed talking about aging, indicating that he thought they would both reach 100.

In spring 1961 Stefansson suffered a stroke, leaving him unable to write, but Leffingwell continued communicating with him through Alan Cooke, a librarian who worked with Stefansson. In one of the letters, Leffingwell conceded that he still carried deep resentment toward Baldwin. But he had also reached the decision that his health was such that he would not be writing anything about Baldwin and the Baldwin-Ziegler Expedition. Leffingwell struggled with emphysema from years of smoking and it was slowing him down.

Stefansson passed away in late August 1962. Leffingwell wrote to Mrs. Evelyn Stefansson in early October and expressed his feelings:

"He was kind enough to say that his memories of me were pleasant. My memories of him are certainly so. As I grew older in years and worldly wisdom, my regard for Stef grew also."[3] She responded in kind to his letter by writing a week later: "He admired you enormously, and in recent years especially enjoyed his correspondence with you, and reminiscing about the old days."[4]

With Stefansson's passing, a door had closed. No one else really understood how profound their shared Arctic experiences had been.

Leffingwell and Anna still took daily walks. He enjoyed family visits and the abundance of flowers in their yard. In 1966, a day before he turned age 91, Leffingwell wrote the following poem for Anna:

> January Twelfth!
> Just fifty years ago today
> E. DE K. L. married Anna May
> And though the wintry winds did blow
> In Satan [*sic*] boots—through all the snow
> As a lovely bride, all dressed in white
> Her mien demure, her eyes so bright.
> There at the church, Ernest awaited.
> An anxious moment, his breath abated
> This handsome pair were there made one
> And a long happy life together begun.[5]

Whatever their differences over the years, they had resolved them.

At age 92, Leffingwell was interviewed by the Monterey Peninsula newspaper. The interviewer wrote: "the one experience he remembers as a 'life or death' decision, when a dory broke its moorings and was drifting away, forcing him to dive into icy water to retrieve it. He and an Eskimo companion cut blocks to build a snow house, in which he thawed out while additional snow was melted and boiled on a seal oil stove, so that scalding tea could ward off pneumonia."[6]

Leffingwell passed away in Carmel on January 27, 1971, just two weeks after his 96th birthday, and several years after his last sibling had passed. He was the last surviving charter member of The Explorers Club of New York.[7] In a letter to Gil Mull, Anna indicated, "He was active and hearty until the last few months."[8] A permit for "Disposition of Cremated Remains at Sea" was obtained from the California Department of Public Health and at 10:30 a.m. on March 27, 1971, Leffingwell's

Leffingwell (in his eighties), in the 1960s in Carmel, California. *Deborah Storre Collection*

ashes were scattered from an airplane three miles west of Point Lobos, California.[9] Anna received condolences from many friends and family, as well as the president of Trinity College, The Explorers Club, and the National Rifle Association. One of the most heartwarming condolences was from a nephew: "As a child, MY uncle Ernest was the man of everything. he took the time to write a letter to ME at age ten which I still have and this adoration has never diminished. he was a man. and my father with all of his bicepts [*sic*] would not mess with him. and he was so very nice to my mother and told her that she was beautiful—a child has far keener perceptions than an adult but this has never changed about MY uncle."[10]

Deborah Storre commented that her grandfather was "not a person who lived in the past" and that he talked about current projects. She has no memory of him ever talking about his time in the Arctic. Deborah describes him as having a "delightful sense of humor" and that he was "kind and generous." She described one of his routines: "From his favorite chair in Carmel, he could see the sun rise in the morning. He prepared a drawing that showed what he could see from his chair and marked the point at which the sun rose each morning. He would follow the sun as it tracked north and then south each year."[11]

The structure he had always known was evident in his later years. Terry Leffingwell Grebens, Eric's eldest daughter, remembered her grandfather as spry, fun, and entertaining. But, she wrote, he was "very prompt," held to a regular schedule, and if the family had not arrived by a particular time, meals began without them.[12] When he passed, his wife Anna, who would survive her husband by four and a half years, told a Monterey newspaper: "He never should have left the Arctic."[13] She understood better than anyone how powerful and profound his experiences in the Arctic had been. Perhaps it partially explains why he chose a life that focused on being outdoors, a life where he was always making something using his hands, whether a house, boat, car, or growing flowers and vegetables. His thirst for knowledge, learning, and his careful attention to detail never waned.

Photographers and writers are able to convey a sense of the Arctic through their images and descriptions. For those who have experienced and found a connection to the Arctic's beauty, harshness, softness, and vast landscape, through efforts by photographers or writers, or by visiting the Arctic Refuge, it is safe to say that we understand his appreciation of the Arctic in a way that others may not. During summertime, there are the cloudless sunny days where the sun never sets but travels in circles above your head. There are the days in mid-summer when it snows. There are amazing landscapes of beauty and constantly changing light

Caribou herd near the Canning River delta, 2008. *Janet Collins Photographic Collection*

on the land. Flowers explode on the tundra in July. Caribou migrate in thousands. Large wolves and grizzly bears wander the region, and are curious or bold or aggressive or wary. Birds everywhere, some squawking as they protect their nests from predators. After all, they have traveled amazing distances to nest in the Arctic. And there are the winters with consistent below-zero temperatures, blowing snow, incessant and piercing winds, and frequent fogs. It is indeed a powerful place.

John Kauffmann wrote in *Alaska's Brooks Range*, "Leffingwell is an unsung hero of Alaskan Exploration."[14] More than one hundred years later, it is long overdue that he receives the recognition and consideration for his scientific work and contributions to Arctic exploration and mapping.

CHAPTER 14

Recognition

"An Arctic explorer after all"

Leffingwell was not the type to boldly promote himself or his work, but he was disappointed not to receive more recognition. His work in the Arctic between 1906 and 1914 coincided with major events in polar exploration. Those events included the "first" to the North Pole, South Pole, and the first through the Northwest Passage. The events overshadowed his work, and recognition of Leffingwell's efforts came slowly. He lived long enough to experience some recognition, but additional notice followed his passing.

Still Leffingwell did not hide his opinions about Arctic exploration. He felt that hardship experienced while traveling in the Arctic was a result of incompetence or one "who does not know how to take care of himself." But Leffingwell had experienced hardships in the Arctic, too. And in a statement open to various interpretations, he indicated that people would not read a book about Arctic exploration without hardship; drama would sell more and benefit the writer. He had written a scientific publication, not a "popular" book. Leffingwell adamantly opposed exaggerating events for dramatic effect or revenue.

A week after his January 1916 marriage, a news article quoted him as saying: "If you are doing something in the interest of science you have to pay your own expenses. I have invested about $30,000 advanced me by my father as my interest in the estate and I have a hope that the world at large will benefit by my investigations but scientific books do not sell and I am not going to write a book for general circulation."[1]

Hudson Stuck, who served as Archdeacon of the Yukon for the Episcopal Church, traveled along the northern Alaskan coastline in 1918 and became aware of Leffingwell's accomplishments. He wrote the Royal Geographical Society (RGS) and proposed that they recognize Leffingwell for his scientific work—in particular, the mapping of the coastline from Barrow to the Yukon border. The RGS response was to award Leffingwell the Patron's Medal in 1922. The medal was presented on May 29, 1922, at the Society's Anniversary Meeting, but Leffingwell was unable to attend. Mr. Oliver B. Harriman, First Secretary of the United States Embassy, attended in his place. The medal was then transferred to the embassy in London, forwarded to the U. S. Department of State in Washington, DC, and then sent on to Leffingwell in California.[2]

The *Los Angeles Times* reported on the Royal Geographical Society award on July 16, 1922, and referred to earlier articles written about Leffingwell, noting that many had "scoffed" at his expeditions, "seeing the cruise in the light of the folly of a rich man's son. Others…lauded the spirit of the youthful explorer."[3]

In a draft response by Leffingwell to the Royal Geographic Society: "I am beginning to think that I must be an Arctic Explorer after all."[4] Leffingwell was clearly pleased by the RGS recognition because his mother had been born in England and he felt an affinity with that nation's people.

Stuck had also written the American Geographical Society, which followed suit and in 1922 awarded Leffingwell with the Charles P. Daly Gold Medal, recognizing "valuable or distinguished geographical services or labors." The presentation of the medal took place on March 31, 1923, at the University of California, Southern Branch (today's UCLA), and was covered by the *Pasadena Star News*. The presentation of the award, "for producing the first accurate chart of a part of the Arctic coast of Alaska and for sustained and original investigations in Arctic physiography,"[5] was followed by an informal talk by Leffingwell for faculty and students, and a tea hosted by the geography department.[6]

There were other awards too. In a letter to Leffingwell dated November 20, 1922, The Explorers Club awarded him with "corresponding membership," noting his "eminent services to science and exploration."[7] Leffingwell's alma mater, Trinity College, awarded him with an honorary degree of Doctor of Science on May 22, 1923, and recognized him at their graduation on June 11.

There are also geographic features named for Leffingwell in the Arctic. Mikkelsen named Leffingwell Nunatak (a rocky peak) after him in northeastern Greenland. Stefansson gave the name Leffingwell Crags to a cliff on Melville Island in the Northwest Territories of Canada.

In December 1972 Hillard Reiser, a geologist with the U.S. Geological Survey in Menlo Park, California, wrote Anna that he had submitted recommendations for features to be named after Leffingwell in the Arctic National Wildlife Refuge. One of them was "the naming of a mountain located near Lake Peters and Schrader in the Mt. Michelson quadrangle, and a mountain ridge and fork of the Aichilik River in the Demarcation quadrangle...It is my hope and belief these will be accepted by the Board of Geographic Names."[8] Today the U.S. Board on Geographic Names recognizes three features in honor of Leffingwell: Leffingwell Creek, Leffingwell Fork, and Leffingwell Glacier. All fall within the Arctic National Wildlife Refuge boundaries. Leffingwell Glacier trends northeast and is located on the south and east side of Mount Michelson. Leffingwell Creek flows out of the glacier into the Okpilak River. Both were identified as such by E.G. Sable in his 1965 U.S. Geological Survey Open File Report 810. Leffingwell Fork originates in the Romanzof Mountains, travels north by northwest and flows into the Aichilik River.

Gil Mull was instrumental in having Leffingwell's Flaxman Island base camp designated as a State Historical Site. He wrote Alaska Governor Egan and Ted Smith, director of the Division of Parks, in the spring of 1970 about preserving the site. Mull also wrote Leffingwell in the spring of 1971, not knowing that he had passed in January: "Your early exploratory and geological work in northern Alaska has now become an important part of Alaskan history, even more so now since the discovery of oil at Prudhoe Bay in the Sadlerochit Formation which you named. As you may have heard, the Prudhoe Bay oil field is by far the largest oil field ever discovered in North America. And the development of this field has strongly focussed attention on the permafrost or ground ice that you studied so comprehensively in that area."[9]

Mull placed a plaque at the site in the summer of 1971 that reads: "From this base camp Geologist Ernest D.K. Leffingwell almost single handedly mapped Alaska's Arctic Coast during the years 1907–1914. He also identified the Sadlerochit Formation—main reservoir of the Prudhoe Bay Oil Field."[10]

The site is owned by the State of Alaska, Division of Land and Water Management. It was recognized in the Alaska Heritage Resource Study dated January 15, 1971, and was listed on the National Register of Historic Places in July 1971 as "Leffingwell Camp State Historic Site." It was designated as a National Historic Landmark in 1978, and is administered by the U.S. National Park Service.

The National Park Service has indicated that there have been a number of alterations to the site. The National Historic Landmark designation can be removed if the site is significantly altered to the extent that it no longer represents the reason it was originally designated as such. The original cabin was dismantled by an early trader and moved to Brownlow Point. The Panningona and Okomailik families built houses on the Flaxman Island site in 1924 and 1940 respectively. However, some of the original items from the Duchess of Bedford remained in 2008 when the author visited the site. They include a cabin from the ship, a boiler, the cellar, and some canvas and metal remnants.

An Inupiaq allotment claim has been submitted and at the time of publication was pending. If the site is re-conveyed, the owner inherits the National Historic Landmark designation.

Sadly, the entire region, including the south coast of Flaxman Island, has suffered from severe erosion due to coastal processes and exacerbated by climate change. Unless preserved, what remains of the Leffingwell site will eventually disappear, as the astronomical pier and other remnants already have.

Gil Mull, who has long appreciated and understood Leffingwell's contributions to scientific knowledge, recently wrote: "His drive and motivation were incredible, and his mapping and delineation of the geologic structure and stratigraphy was damn good. We who came along behind have added many refinements, but there is little of his broad regional framework that was just flat wrong."[11]

Coastal erosion on Barter Island, 2008. *Janet Collins Photographic Collection*

Epilogue

The Arctic fed Leffingwell's soul. Though he never returned to the Arctic, his experiences there remained in his heart. Given his passion for the Arctic, why did he not return? By the time USGS Professional Paper 109 was published in 1919, Leffingwell had married and moved on with his life. He had responsibilities and his family to think of, and his resources were limited.

Pursuing his early interest in Arctic exploration set Leffingwell off as different from the rest of his family. Though he was not business-oriented, he knew how to work hard. Still, he may have felt pressure to conform to societal expectations. He once indicated to Anna that not everyone in the family supported his work in the Arctic. But he also questioned himself, believing that he had not produced much during the time he was up north. In fact, he produced an amazing amount of work considering the challenges he faced. The assistance of the Inupiat was critical to his efforts and success, and he recognized the importance of their help.

His exploration and mapping in the Arctic was a result of gaining the necessary skills and experience to survive and flourish in a harsh environment, and combining that with determination, persistence, a passion for science and the Arctic, resiliency, and discipline. He faced many obstacles along the way and overcame most of them. Leffingwell had high expectations of others and higher ones for himself. He was driven by science, but showed flexibility as plans and situations constantly changed and were in a state of flux. One might also wonder why Leffingwell never pursued a career in geology—or why he has not been well known. He declined offers to work for the Geological Survey or in academia. When offered positions, he used the low salaries and his lack of continued studies as reasons to decline. Yet Leffingwell was a renaissance man, interested in anything and everything, and very capable, whether it meant building a boat, a car, or a house. And he never stopped learning. He bored easily and needed creative outlets. It would be hard to imagine Leffingwell sitting at a desk; he had a powerful connection with nature and loved being outdoors. Perhaps the best answer is that traditional societal expectations could never define him.

So why has fame bypassed him? Leffingwell's efforts were greatly overshadowed by other explorers and expeditions, and he was not one for self promotion. He was described in one interview as "unassuming and somewhat shy." His father was the one who submitted articles to the press and championed his son's work. In contrast to Stefansson, Leffingwell did not write books. He wrote a scientific publication that simply could not generate the same kind of revenue and publicity as popular reading.

Another question that comes to mind given the controversy surrounding development in the Arctic Refuge. What would he have thought of the oil and gas development on the North Slope? I have no idea, although in an interview Leffingwell made reference to President Harding setting aside from development the Petroleum Reserve Number 4, now known as the National Petroleum Reserve. Leffingwell commented, "may his tribe increase." His wife was a founder of the Sierra Club chapter in Carmel, and their lives focused on the outdoors and nature. Their son Eric became a state park ranger. Before Leffingwell died, a group of U.S. Geological Survey geologists interviewed him and taped the conversation, but the tape has disappeared. A better sense of his perspective about oil development might have been gleaned from the tape.

Leffingwell's work in Alaska is as relevant today as when he accomplished it. He was motivated by Fridtjof Nansen's 1897 lecture to contribute to scientific knowledge, a goal Leffingwell took very seriously. His determination, passion, resiliency, and meticulous nature served him well. Increasing interest in his work and life would probably surprise and please Leffingwell.

Geologists in particular have tremendous respect for the quality of his work and the difficulties in accomplishing it. If one compares his geology map to more current maps, the similarities are easily found—for example, the faults, the east/west bands of geologic material that define the region, the Sadlerochit Mountains, and the granite of the Romanzof Mountains. His report describes in detail geomorphology, drainages, and physiography. He identified and mapped Paleozoic sedimentary rock and assigned names such as Neruokpuk schist, black shale,

Lisburne limestone, and Sadlerochit sandstone. He detailed Mesozoic rocks and their names: Shublik Formation, Kingak shale, and Ignek Formation. And Leffingwell mapped the igneous rock, both Paleozoic and Mesozoic, and the Flaxman Formation of the Cenozoic. His work provided a foundation for others to build upon. Throughout his many years working on the North Slope, geologist Gil Mull continues to be amazed by Leffingwell and his work. Leffingwell named the Sadlerochit Formation, which is the major reservoir in the Prudhoe Bay Oil Field and according to Gil Mull, "by far the largest ever discovered in North America." Leffingwell's report included details on the oil seepage near Cape Simpson. The Inupiat had been aware of it for some time. Leffingwell wrote: "These oil bearing rocks may also occur in other parts of the Arctic slope region. Even if an oil pool were found in this northern region, there is serious doubt of its availability under present conditions, though it might be regarded as a part of the ultimate oil reserves that would some time be developed."[1]

Leffingwell's pioneering work on ground ice (permafrost) and explanation of ice wedges was critical knowledge for any development wherever ground ice existed, and certainly for oil development on the North Slope of Alaska.

The former U.S. Coast and Geodetic Survey, now the National Ocean Service, many years later reviewed Leffingwell's work in mapping the coastline from Barrow to the Yukon border, and expressed wonder at the accuracy of his work. Leffingwell wrote in his report: "The accuracy of the base net is about that of the secondary triangulation of the United States Coast and Geodetic Survey."[2] The mapping accuracy of latitude was about plus or minus .1 of a second of a degree, and because he used lunar observations, the longitude was accurate to plus or minus 7.0 seconds of a degree. Leffingwell was aware of the drawbacks of lunar observations versus a radio transmitted signal. If one compares his 1:250,000 Reconnaissance Map of the Canning River Region to a current U.S. Geological Survey topographic map, several locations where longitude is inaccurate are quickly noted. For example, the distance between the Sadlerochit River and the Hulahula River are depicted closer than they actually are. And Shublik Island and Shublik Springs are shown closer than the current map indicates. The maps are amazingly accurate, nonetheless.

Ned Rozell, in an article for the University of Alaska *Science Forum* (February 4, 2009) wrote about Leffingwell as "a scientist with a fan club." Among those fans is Dr. Kenji Yoshikawa, a permafrost scientist at the University of Alaska Fairbanks, who wrote:

> Once the USGS professional paper was published in 1919, it was widely known as Leffingwell's work. His USGS professional paper is not just a typical science report or results paper. I believe its 251 pages demonstrate his passion and experiences in the last frontier of Alaska. He is definitely the last classic polar explorer. He focused on science which was unusual for most Arctic expeditions of that time.
>
> When I think about Leffingwell's expedition, there are several remarkable points that should not be forgotten. He did an amazing review of the literature about permafrost. He checked Russian and German literature to understand what the knowledge of ground ice (or frozen ground) was at that time. Many of the periglacial processes were not well understood in the early 20th century. However, Leffingwell used correct logic and figured out problems based on his year-round detailed observations, especially regarding ground ice and coastal erosion.
>
> Today, we could learn many things from him, but I will focus on three points...: 1) motivation will follow passion; 2) detailed observation, knowledge (his review of the literature), and motivation will reach new frontiers, and 3) Leffingwell's many survey points for his mapping work that included pingos and coastal locations will be very valuable information for future changing Arctic regions. It is unique information for the rest of the Arctic. The results of his work will forever remain and will continue to contribute to the scientific community and his passion for his work will continue to the next generation. Because science without passion is like life without dreams.[3]

Another fan is Dr. Matt Nolan, a scientist at the University of Alaska Fairbanks, who studies climate change and glaciers. By comparing Leffingwell's 1907 photographs of the Okpilak Glacier to current day photos, Dr. Nolan has been able to document the retreat of that glacier over the past century.

Then there is Joe Henderson, who lived on the North Slope for many years and recounted his personal sledging adventures as he retraced Leffingwell's travels in *Mushing Magazine*.

It is important to recognize and honor the Inupiat who assisted Leffingwell and made his accomplishments possible. Leffingwell became close to, worked with, and depended upon Inupiat living on the North Slope. Out of respect for him, the Ericklook and Shingatok families each named sons Leffingwell.

During the 1960s Leffingwell donated a number of his books to the Carmel Public Library. They were eventually transferred to the historical collection of the Alaska State Library in Juneau. He also donated his papers to the Stefansson Collection at Dartmouth Library.

The Smithsonian Museum has four primary collections of items provided by Leffingwell and his efforts in northeastern Alaska: 97 specimens categorized under anthropology, 17 under fossils, 39 under mammals, and an unspecified number of geology specimens.

Leffingwell left behind an impressive legacy that might have surprised him most of all. He, along with Mikkelsen, traveled to the Arctic in 1906 and resolved a longtime controversy about whether there was land north of Alaska by determining that the edge of the continental shelf sits relatively close to the Alaskan coastline. Later he mapped the coastline accurately between Barrow and the Yukon border, conducted foundation studies of ground ice (permafrost), and defined the geography and geology through nomenclature and mapping of a significant portion of the Arctic National Wildlife Refuge. His final report for the U.S. Geological Survey detailed a composite of his knowledge of survival in the Arctic, including equipment, food, shelter, and clothing, and represents the extent to which the Arctic impacted his life.

Anna Meany Leffingwell, his wife of 55 years, may have expressed Leffingwell's passion for the north better than anyone: "He never should have left the Arctic."

APPENDIX 1

Expedition Members

I. Baldwin-Ziegler Expedition 1901-1902

Evelyn Briggs Baldwin	Commander
Captain Carl Johanson	Sailing Master
Captain Johan Menander	Chief Officer
Captain Magnus Arnesen	Ice Master
Russell W. Porter	Artist
Anton M. Vedoe	Electrician
H.P. Hartt	Chief Engineer
Axel Ivan Axell	2nd Engineer
Chas. E. Rilliet	Engineer
Ernest Leffingwell	Chief Scientific Officer
Ejnar Mikkelsen-Loth	Geodesist
James P. DeBruler	MD
Chas. L. Seitz	MD
Wm. W. Verner	MD
Anthony Fiala	Photographer
Leon F. Barnard	Secretary

Additional crew members: Hermann Andree, Archibald Dickson, J. Knowles Hare Jr., Carl L. Sandin, Sverdrup J. Vedoe, and Robert L. Vinyard. There were 17 Americans in all, a Swedish sailing crew, and 6 Russian dog caretakers.

II. Anglo-American Polar Expedition 1906-1908

Ernest Leffingwell	Joint Commander
Ejnar Mikkelsen	Joint Commander
George Howe	MD
Ejnar Ditlevsen	Botanist, left expedition at Teller due to indigestion and acute pain
Mr. Edwards	Mate, left expedition at Teller due to malaria
Storker Storkerson	Mate
Christopher Thuesen	Mate
J. Parker	Mate, left expedition at Teller
Cook	left expedition at Teller
Joe McAlister	Mate, transferred from the *Thetis*
Max Fiedler	Mate, transferred from the *Thetis*
William Hickey	Mate, transferred from the *Thetis*
Joe Carroll	Cook, came aboard at Barrow
Vilhjalmur Stefansson	Ethnologist, traveled via Mackenzie River to Herschel Island

APPENDIX 2

Timeline

Pre-Alaska

1875	Leffingwell born January 13, Knoxville, Illinois
1895	graduated Trinity College
1895-1896	taught at St. Alban's School, Knoxville
1896-1906	attended University of Chicago
1898	U.S. Navy—Spanish American War
1901-1902	Baldwin Ziegler Expedition
1903	Summer—geology field work, Wyoming
1903-1904	Superintendent at St. Alban's School, Knoxville
1904	Summer—geology field work, Colorado
1905	Summer—geology field work, Washington State

Alaska

1906	Anglo-American Polar Expedition
	May 20—departed Victoria, BC, Canada
	September 17—reached Flaxman Island
1907	March 17—pack ice trip
	May 17—Okpilak River with Ned and Gallagher Arey
	July 30—mapped a few square miles near Collinson Point
	August 5—to Flaxman Island
	October 27—Canning River with Okalishuk
1908	January 23—Hulahula River with Inupiat Flavina and Mamayuok
	March 31—Canning River
	May 6—Canning River to Ignek Creek with Okalishuk
	July 15—mapped mouth of Canning, with Okalishuk and Toxiak
	August 17—to Barrow on Karluk
	September 8—returned south on *Narwhal*
1909	May 20—departed Seattle on the *Argo* for Flaxman Island
	August 23—reached Flaxman Island
	August 28—trip to Barrow for supplies
	Fall—no field work, built new house by old one, constructed astronomical pier

1910	April 1—to Point Barrow for whaling
	July—mapped near Oliktok, reached Flaxman Island, mapped Brownlow Pt. S.E.
	August 1—to Barrow for provisions and mail on *Argo*
	August 31—mapped 50 miles west of Flaxman with McIntyre
	Fall—no field work, banked house with snow, made tents, sleds
1911	February 11—to Canning River and north front of Franklin Mountains
	April 2—sled trip west to Oliktok, continued establishing triangulation network
	May 10—to Sadlerochit Mountains to Peters and Schrader Lakes with Aigukuk and Karoiga
	late June and July—mapped coastal topography west of Gwydyr Bay
	August—to Barrow on *Argo*, 2 weeks mapping Flaxman Island
	September—mapped east mouth of Sagavanirktok
	Fall—no field work, repaired the *Argo*
1912	January and February—latitude observations at Flaxman Island
	March—established triangulation stations to west, Sagavanirktok
	April 10 and May—sledded west to Oliktok, with Karoiga and Nanjek, completed most of triangulation network
	June 6—mapped coast to Barter Island with Karoiga and likely Aigukuk
	August 3—reached Flaxman house, turned *Argo* over to McIntyre
	August 15—left Flaxman with Inupiaq family for Barrow
	September 2—reached Point Barrow
	late September—headed south aboard the *Belvedere*
	early November—reached San Francisco
1913	June 24—traveled Seattle to Nome on the passenger ship *Senator*, met Stefansson and CAE, joined supply ship *Mary Sachs* for transport to Flaxman Island
	August 15—reached Barrow
	August 18—reached Flaxman Island and decided to over winter
	October—adjusted base triangulation network
	December and January—with Canadian Arctic Expedition Southern Party at Collinson Point
1914	February 14—to Challenge Island with Terigloo's 13-year-old son
	March 8—near mouth of Kadleroshilik and Kuparuk Rivers
	April 21—arrived Flaxman
	late April–early May—traveled 20 miles up the Canning River
	May—three weeks at Flaxman to finish work and pack
	May 20—to Collinson Point and CAE, one month at Katakturuk River

July 1—reached Flaxman Island
July 11—left Flaxman Island for Barrow
August 17—reached Barrow
August 21—aboard the *Jeanette*, reached San Francisco 6 weeks later

Post Arctic

1915	worked on U.S. Geological Survey, Professional Paper 109 at USGS, Washington, DC, met Anna Meany of Chicago
1916	January 12, married Anna Meany (b. July 2, 1890)
1917	December 3, daughter Ann [Nancy] born
1926	November 6, Leffingwell's mother (b. May 5, 1840) passed
1928	October 9, Leffingwell's father (b. December 5, 1840) passed
1930	June 24, adopted Eric (b. Aug 20, 1928) and Christine (b. November 13, 1929)
1937, 1939	worked in Imperial Valley
1940	acquired Rocky Creek property south of Carmel
1946	Rocky Creek property relinquished
1947	acquired Klamath River property and built house
1955	second flood washed away Klamath house
1971	January 27, Leffingwell passed
1975	June 27, Anna passed
1986	November 22, Christine passed
1997	November 14, Eric passed
2001	July 1, Ann [Nancy] passed

APPENDIX 3

Okpilak Sledge Trip, May–July 1907

Leffingwell with Ned and Gallagher Arey

Equipment

4 dogs, dogs etc.
4 dog pack saddles
Camp Items:
wall tent 7 x 8
stove, cached below mts.
Primus & cleaner
16 lbs oil & funnel
1 large iron pot
1 smaller tin pot
2 fry pans small
3 bowls
3 plates
3 spoons
1 skin knife
4 squirrel traps
303 Rifle Savage
150 rounds
larger pistol
100 rounds
Personal Items:
2 pipes & pouch (French novel)
3 lbs plug tobac
watch
thermometer
file
sheath knife
twine
small cord
small wire
whetstone
fountain pen
ink powder
scissors
2 needles
sewing gear
blanket pins
mirror
vaseline
Shakespeare
blanket single 5 lbs
sheep skin 5 lbs
Clothing Items:
handkerchief
1 suit underwear
flannel shirt
fur shirt-lost June 17
overalls & belt
nob shoes
rubber shoes not used
moccasins
water boots
socks 3 pr
snowshirt cached
Pack sack
ditty bag
towel & soap
mosquito net
mosquito dope
Surveying Items:
plane table needles
base bar
30 sheets in case
notebook & case
2 hard pencils
3 soft pencils
1 red & blue pencil
2 rubber
aneroid
2 triangles
1 ruler scale
protractor
az compass
therm tad.
draw pen
india ink solid
pocket tape

Food

3 lb salt in tins
8 lbs biscuit
1/4 pepper in tins
5 lard
40 sugar

50 cornmeal
50 rice
25 granola
25 butter
12 tins jam

<4 tea
dried vegetables
12 tablets
3 qts whiskey

Source: MSS 69, Box 1, Folder 13, Dartmouth College Library, Rauner Special Collections.

APPENDIX 4

Selected Geographic Names

The following list is an abridged version of Leffingwell's List of Geographic Names in the "Geographic Nomenclature" chapter of USGS Professional Paper 109, *The Canning River Region, Northern Alaska*, pages 93–100. Leffingwell explained, "With the·exceptions here mentioned all the names have been traced to their original sources…In the following catalogue the names in bold-faced type have been entered on the writer's map. The authority comes next after the place name, and then the source of the name—where it has been ascertained."

Aichillik; river. Leffingwell. Empties into the Arctic Ocean near the 142d meridian. Spelled Aitshillik by [Canadian ethnologist Diamond] Jenness. Meaning uncertain.

Arey; island: Leffingwell. After the prospector, H.T. Arey, who lived in the neighborhood several years.

Barter; island: Franklin, Leffingwell…This island was named Point Manning by Franklin, but local usage has changed the name. Franklin's Barter Island is called Arey Island on the writer's map…

Chamberlin; mountain. Leffingwell. After Prof. T. C. Chamberlin. A conspicuous peak, 9,000 feet high, at the headwaters of the Sadlerochit River.

Ignek; creek. Leffingwell. The first tributary on the east side of the Canning. Means "fire." Jenness spells it Ignik.

Itkilyariak; creek. Leffingwell. A tributary to the Sadlerochit. Means "route by which the Itkillik travel."

Jago; river. After Lieut. Jago, of Collinson's ship. Empties into the Arctic Ocean at Martin Point, near the 143d meridian. The Eskimo name is Jags.

Katakturuk; river. Leffingwell. Empties into Camden Bay. Arctic coast. Perhaps means "a narrow place" (?).

Konganevik; point. Leffingwell. A point on the west side of Camden Bay, Arctic coast. Jenness spells it Kongangevik. Means "place where there is a deer pond."

Michelson; mountain. After Prof. A. A. Michelson. A glacier-clad peak on the west side of Okpilak River.

Okpilak; river. Leffingwell. Enters the Arctic Ocean at the 144th meridian, having a delta in common with the Hulahula River. Means "no willows."

Peters; lake. Leffingwell. After W. J. Peters, of the United States Geological Survey. One of two lakes on the headwaters of Sadlerochit River. Eskimo: Neruokpuk, "big lake." Jenness spells it Narivukpuk.

Red; hill. Leffingwell. At the west end of the Sadlerochit Mountains.

Sadlerochit; mountains. Leffingwell. The northern front range of the Arctic Mountains between Sadlerochit and Canning rivers.

Sadlerochit; river. Leffingwell. Empties into the Arctic Ocean near longitude 144°30'. Means "the area outside of the mountains" (?). Probably the same as Marsh's Barter River.

Salisbury; Leffingwell. mountain. After Prof. R.D. Salisbury. A conspicuous double snow-clad peak a few miles west of Canning River.

Schrader; lake. Leffingwell. After F. C. Schrader, of the United States Geological Survey. One of two lakes on the headwaters of Sadlerochit River. Eskimo: Neruokpuk, "big lake."

Shublik; Leffingwell. mountains. An outlying chain of mountains, between Canning and Sadlerochit Rivers.

Shublik; springs. Leffingwell. General locality near a large spring on [the] Canning River. Means "a spring." Jenness spells it Sublik.

Tamayariak; creek. Leffingwell. A tributary which enters the Canning from the east, near the coast. Means "route where some people were lost."

Weller; mountain. Leffingwell. After Prof. Stuart Weller. A peak near the east end of Sadlerochit Mountains.

Additional Place Names [from Ernest Leffingwell's handwritten notes]

Itkilyariak = Indian route.
Kaktoavik = seining place.
Kangingnvik = place for deer drive.
Katakturak from Katak = gap in mts.
Kingak = nose or steep end of ridge.
Kugruak = Old river.
Oliktok = it trembles shakes / caribou.
Tamayariak = Rout on which one got lost.
Tigvariak = the way, place, rout, etc. to go to land.

Source: U.S. Geological Survey, Anchorage, AK. Leffingwell Collection. Book 356, *Northern Alaska, Canning River Region, 1906-1914.*

APPENDIX 5

Celsius to Fahrenheit Conversion Table

°C	°F	°C	°F	°C	°F	°C	°F
50	122.0	27	80.6	4	39.2	-19	-2.2
49	120.2	26	78.8	3	37.4	-20	-4.0
48	118.4	25	77.0	2	35.6	-21	-5.8
47	116.6	24	75.2	1	33.8	-22	-7.6
46	114.8	23	73.4	0	32.0	-23	-9.4
45	113.0	22	71.6	-1	30.2	-24	-11.2
44	111.2	21	69.8	-2	28.4	-26	-14.8
43	109.4	20	68.0	-3	26.6	-27	-16.6
42	107.6	19	66.2	-4	24.8	-28	-18.4
41	105.8	18	64.4	-5	23.0	-29	-20.2
40	104.0	17	62.6	-6	21.2	-30	-22.0
39	102.2	16	60.8	-7	19.4	-31	-23.8
38	100.4	15	59.0	-8	17.6	-32	-25.6
37	98.6	14	57.2	-9	15.8	-33	-27.4
36	96.8	13	55.4	-10	14.0	-34	-29.2
35	95.0	12	53.6	-11	12.2	-35	-31.0
34	93.2	11	51.8	-12	10.4	-36	-32.8
33	91.4	10	50.0	-13	8.6	-37	-34.6
32	89.6	9	48.2	-14	6.8	-38	-36.4
31	87.8	8	46.4	-15	5.0	-39	-38.2
30	86.0	7	44.6	-16	3.2	-40	-40.0
29	84.2	6	42.8	-17	1.4	-41	-41.8
28	82.4	5	41.0	-18	-0.4		

Source: National Weather Service, National Oceanic and Atmospheric Administration, www.srh.noaa.gov/ama/?n=conversions#tables.

Acknowledgments

Following my presentation of a paper on Leffingwell at the Western Association of Map Libraries (WAML) conference in Alaska in 2005, I shared dinner with Barbara and David Haner and Russell Guy. They encouraged me to write a book about Leffingwell. I am grateful to them. Barbara passed away in November 2010, but she remains in my heart and thoughts. WAML was the foundation and the support system of my career in map librarianship and I owe a tremendous amount to map librarians.

As I began the research journey, one of my college roommates, author and librarian Rae Jensen McDonald, offered encouragement and a positive perspective. My gratitude extends to renowned author John Taliaferro, whom I approached while he was touring for his book *In a Far Country*. I later wrote him presenting my interest in writing about Leffingwell. His kind words, support, encouragement, and pep talk were critical in moving forward with the project. Don Grybeck kindly provided me with a copy of Leffingwell's work, U.S. Geological Survey Professional Paper 109. Sadly, he passed away before seeing this book published.

As I delved into the research I had accumulated, I gained a profound appreciation for the role of the American Geographical Society and the Royal Geographical Society in supporting the dreams of explorers. That appreciation carries to the current day. Peter Lewis, librarian, and Maria Victoria Rosa at the American Geographical Society in New York City provided access to their files, and council meeting minutes. The Royal Geographical Society in London provided significant background information on the funding of the Anglo-American Polar Expedition and their efforts in the eventual recognition of Leffingwell's work on the North Slope of Alaska. Karen-Margrethe Bruun kindly helped with locating specific resources in Denmark. Bent Nielsen and Annika Egils-dottir Hansen of the Danish Arctic Institute were very kind, helpful, and resourceful as were the folks at the Danish Royal National Library.

Appreciation goes to Dartmouth College Rauner Library Special Collections, the Manuscript Division of the Library of Congress (LC), Ed Redmond of the LC Geography and Map Division, Barbara Natanson and Kristi Finefield of the LC Prints and Photographic Division, Sierra Laddusaw for her help with a Franz Josef Land map, the Smithsonian Archives, Dorothea Sartain, Lacey Flint of The Explorer's Club, and the University of Chicago Special Collections. Thanks also to Mary McAndrew, Archives and Manuscripts at Knox College, Illinois; Peter Knapp, Special Collections at Trinity College, Connecticut; Sally Hutchcroft and Pat Rose of the Knoxville Public Library at Knoxville, Illinois; Karen Sandoval, who inadvertently acquired some Leffingwell materials and returned them to his descendants; Monterey County, California, offices of the recorder, assessors, and public works; Dennis Copeland of the Monterey Public Library; and Lynn Housouer, Archivist at the Pioneer Museum and the Imperial County Historical Society in Imperial, California. I also visited Del Norte County, California, offices and the County Museum where staff photocopied articles and answered questions about the Klamath River floods of 1953 and 1955. The Victoria City Archives and staff at Flagstaff Public Library were very helpful in obtaining interlibrary loan books. Additional thanks to map librarians, some now retired: Lori Sugden Fuglem, Julie Hoff, John Kawula, Mary Douglass, Matt Parsons, and Evan Thornberry.

The Alaska contingency includes Jill Schneider and Orion Newell at USGS in Anchorage, Rose Speranza at the University of Alaska Fairbanks (UAF) Special Collections, and Matt Nolan and Kenji Yoshikawa, UAF faculty. Ken Pratt led me to Janet Clemens of the National Park Service in Anchorage who was particularly helpful in my understanding of the National Historic Landmark designation and the Flaxman Island site. Many thanks to Kirk Sweetsir, bush pilot, for sharing his knowledge, perspective, and meaningful conversations, along with Robert and Jane Thompson, Sarah Campbell, and her son Kevin. And thanks to Scott McGee, cartographer, Realty Division, U.S. Fish & Wildlife Service, for his beautiful map of the Arctic National Wildlife Refuge.

The folks at the National Archives of Canada were great. They retrieved diaries of selected members of the Canadian Arctic Expedition for me to gain additional insight into other's opinions of Leffingwell. Leffingwell had several lengthy visits with members of the southern party at Collinson Point and collaborated extensively with several of them. And thanks to Benoit Theriault and Jonathan Wise of the Canadian Museum of History in Ottawa for their assistance.

Shayne Vacher, Tamara Becker, and Denny Vaughan helped with insight into the nautical world and shared some unforgettable sailing on Puget Sound. Gene Hoerauf, cartographer, GIS whiz, friend, and former colleague, has always been amazing and helpful. One of his maps provides meaning and understanding to the narrative. I met another cartographer and GIS whiz, Dan Bowles (an Aussie), on a 2012 trip through the Northwest Passage in the high Canadian Arctic. Dan took an immediate interest in the project and was able to read the proposal and the first few chapters. On the same trip I met historian and author David Pelly, who offered encouragement and shared his wisdom and knowledge. My sister and family genealogist, Sara Grilley, generously researched the Leffingwell family, identifying his siblings and locating grandchildren. Two of Ernest Leffingwell's grandchildren have been another wonderful resource: Deborah Storre and Terry Grebens. Terry led me to Clive Rayne (recently deceased), who provided additional insight about Leffingwell, and was a friend of her father's, Eric Leffingwell. Kent Sturgis kindly reviewed the manuscript and offered invaluable feedback on all aspects. Gene Myers, former colleague and longtime friend, offered thoughtful and creative approaches to all aspects of the entire project. I owe all a debt of gratitude.

However, my largest debt goes to my readers for the project. Their ongoing and relentless interest, encouragement, support, and our endless conversations explain why the book is being published. As a North Slope geologist who is interested in history as well, Gil Mull knows a tremendous amount about the area and Leffingwell. He was an early "Leffingwell groupie" and instrumental in having the Flaxman Island site initially designated as a State Historic Site. He has consistently and kindly shared his knowledge, opinions, and suggestions. Others who have shared their wisdom include former colleagues and longtime friends: Donna Koepp, a retired documents and map librarian, Peter Stark, former map librarian, Katy Velasquez, Jean Melious, and James W. Scott (British born historical geographer), who passed away in July 2011 but read the Baldwin-Ziegler chapters and the book proposal. Jim is sorely missed, and offered the guidance and wisdom that only an eighty-six-year-old with extensive knowledge of history, writing, and publishing could.

Thanks and appreciation for their friendship and support: Rhonda Smith, Muriel Strickland, Lundee Amos, and many Bellingham friends, including Tim Hostetler and Janet McLeod, Kay Reddell, Katy Batchelor, Myra Ingmanson, and Andy Ross. As I researched throughout the United States, many friends provided me with a place to stay including Carole and Paul McGuire, Peter Stark and Sheila Klos, Phil Perkins (deceased), Sylvia Bender, and Eric and Tricia Wolber. I must also express my appreciation to Tom Storch for his administrative and managerial skills; a rarity in certain working environments. He made a tremendous difference in my life when it was particularly difficult.

I also owe a debt of gratitude to Richard G. Smith (deceased), a geographer who taught an Arctic Environment course that sparked my fascination with Arctic geography and history.

My father, Frank, now deceased, and his wife were interested in the project and provided lodging for six weeks in the Midwest, as I came and went researching in 2009. The farm that my father grew up on and had returned to, is about one and one-half hours by car from Knoxville, Illinois, where Leffingwell grew up. My appreciation extends to my sisters and brothers, Sara, Judy, Wayne, and Dave, and my "third" brother, Lee, for their love and support. Also to Paul, Wendy, Rachel, and Daniel Grilley.

And finally, I am eternally grateful to the folks at Washington State University Press, especially Robert Clark, Beth DeWeese, Caryn Lawton, and Kerry Darnall. Thank you for your belief in my project, your efforts on my behalf, and your kind, supportive encouragement. I feel so lucky to have found you.

Notes

Introduction

1. U.S. Geological Survey Geographic Names Information System, based on the National Elevation Dataset, geonames.usgs.gov. Recent mapping work by Dr. Matt Nolan has determined the elevation of Mount Isto, 8,975.1 feet; Mount Hubley, 8,916 feet; Mount Chamberlin, 8,898.6 feet; and Mount Michelson, 8,852 feet. fws.gov/refuge/arctic.

2. Flaxman Island, located in the Beaufort Sea east of Prudhoe Bay, is 58 miles west of Barter Island, and three miles offshore northwest of the Canning River Delta.

3. U.S. Fish & Wildlife Service, fws.gov/refuge/arctic/purposes.html.

Prologue

1. "Nansen Tells of His Trip," *Chicago Daily Tribune*, November 18, 1897, 1–2. archives.chicagotribune.com/1897/11/18/page/1/article/dr-nansen-tells-of-his-trip.

Chapter 1

1. Container 1, Evelyn Briggs Baldwin Papers, Manuscript Division, Library of Congress, Washington, DC. Hereafter "Baldwin Papers."

2. John Edwards Caswell, *Arctic Frontiers: United States Explorations in the Far North* (Norman: University of Oklahoma Press), 177.

3. Container 4, Baldwin Papers.

4. According to Trinity College, Leffingwell's master's degree from Trinity was dated 1905, although he started taking classes at the University of Chicago in the fall of 1896. Leffingwell provided differing years for graduation on his various alumni forms. Watkinson Library, Trinity College, Hartford, Connecticut. A separate news article indicated that he had earned a master's degree in physics from Trinity in 1899.

5. Correspondence to Gil Mull from Anna Meany Leffingwell, May 20, 1973. Gil Mull Files.

6. Container 8, Baldwin Papers.

7. The Arctic Club was originally referred to as the Peary Arctic Club. Its mission was to "promote and encourage explorations of the Polar regions," collect scientific specimens, and maintain manuscripts of expeditions. "Ten Years of the Peary Arctic Club," *National Geographic Magazine* 19 (1908), 661–62. There was also another Arctic Club, formerly known as the Alaska Club, which was located in Seattle, Washington, 1903–1971.

8. Container 3, Baldwin Papers.

9. In this work, Mikkelsen-Loth is referred to as Mikkelsen and where quoted as Loth will include Mikkelsen in parentheses.

10. Leffingwell Papers, Stefansson Mss-69, Box 1, Folder 11, Dartmouth College Library, Rauner Special Collections. Hereafter, "Leffingwell Papers, Mss-69."

11. Container 8, Baldwin Papers.

12. Leffingwell Papers, Mss-69, Box 1, Folder 11.

13. – 16. Ibid.

17. Mss-69, Box 11, Folder 11.

18. Alice Barnard Thomsen, daughter of Baldwin's cousin Leon Barnard, wrote in a synopsis of Barnard's diaries held by the Library of Congress: "The charted path of the *America* during this period looks like the course of a Staten Island ferryboat. The humor of these trips at last began to appeal to the disgusted explorers, and they christened the trip to Cape Tegetthoff on Hall Island and back, 'the mail route.' The time that was not spent on the 'mail route' was consumed in purposeless unloading and loading the provisions and stores from the *America* and back again into her hold." Container 1, Leon F. Barnard Papers, Manuscript Division, Library of Congress, Washington, DC. Hereafter "Barnard Papers."

19. Leffingwell Papers, Mss-69, Box 1, Folder 11.

20. Container 8, Baldwin Papers.

21. Container 1, Barnard Papers.

22. – 26. Ibid.

27. The telephone was invented in 1876–77, electric motor in 1821, power industry evolved in 1881. Power for the lines would likely have been generated by coal.

28. Leffingwell Papers, Mss-69, Box 1, Folder 11.

29. Container 1, Barnard Papers.

30. Container 2, Baldwin Papers.

31. Container 8, Baldwin Papers.

32. Container 1, Barnard Papers.

33. – 34. Ibid.

35. Container 9, Baldwin Papers.

Chapter 2

1. Container 4, Baldwin Papers.

2. Container 3, Baldwin Papers.

3. Deborah Storre Collection. Baldwin later published Leffingwell's remarks in the *NY Outlook* on September 16, 1905.

4. Container 4, Baldwin Papers.

5. Ibid.

Chapter 3

1. See E. de K. Leffingwell, *The Canning River Region, Northern Alaska,* U.S. Geological Survey, Professional Paper 109 (Washington, DC: Government Printing Office, 1919), 59. Hereafter: "Professional Paper 109."
2. Correspondence from Mikkelsen to AGS. American Geographical Society of New York Archives, American Geographical Society Library, University of Wisconsin-Milwaukee Libraries. Hereafter AGS Archives.
3. The HMS *Investigator* was located in 2010 near Banks Island in northern Canada. The Franklin Expedition, lost in 1845, has been the subject of numerous historical accounts.
4. Ejnar Mikkelsen, *Mirage in the Arctic: Explorations in Unknown Alaska,* introduction by Lawrence Millman (Guilford, CT: Lyons Press Arctic Adventure Classic, 2005), 19.
5. Mikkelsen, *Mirage in the Arctic,* 29.
6. Correspondence from Chandler Robbins, Chairman of Committee, to Mikkelsen, February 19, 1906. AGS Archives.
7. There are discrepancies in various accounts of the total dollars raised.
8. Roosevelt was interested and knowledgeable about exploration. During their conversation Roosevelt asked for Mikkelsen's opinion about the failure of the Baldwin-Ziegler Expedition. Mikkelsen, *Mirage in the Arctic,* 27.
9. Correspondence from AGS Corresponding Secretary to Hon. Henry Cabot Lodge, U.S. Senate, February 23, 1906. AGS Archives.
10. Correspondence from Y. H. Metcalf, Secretary, Department of Commerce and Labor to Hon. William Loeb Jr., Secretary to the President, March 6, 1906. AGS Archives.
11. Mikkelsen, *Mirage in the Arctic,* 31.
12. Leffingwell Papers, Mss-69, Box 1, Folder 12.

13. – 24. Ibid.

Chapter 4

1. Leffingwell Papers, Mss-69, Box 1, Folder 12.
2. The proper Inupiaq name is Shugvichiak, but Leffingwell wrote his name as Sachawachiak or Sagi. Shugvichiak will be the name consistently used in the narrative, except when quoted.
3. Leffingwell, "Flaxman Island, A Glacial Remnant," *The Journal of Geology* 16:1 (Jan–Feb. 1908): 56–63.
4. Leffingwell Papers, Mss-69, Box 1, Folder 12.

5. – 9. Ibid.

10. Each degree of latitude and longitude consists of 60 minutes and each minute consists of 60 seconds. 1.7 seconds was very precise in the early 1900s. One second of a degree equals one nautical mile.

11. Leffingwell Papers, Mss-69, Box 1, Folder 12.

12. Ibid.

13. Mikkelsen, *Conquering the Arctic Ice*, 178.

14. Leffingwell Papers, Mss-69, Box 1, Folder 12.

15. Ibid.

16. Leffingwell Papers, Mss-69, Box 1, Folder 12. Leffingwell was referring to Mount Chamberlin, and Peters and Schrader Lakes, which he later named. He eventually traveled to the area for further mapping.

17. Leffingwell referred to him as Okalisuk, but Okalishuk is the proper Inupiaq name and will be used throughout the book.

18. Leffingwell's account says Fiedler went along, while Mikkelsen's did not. Leffingwell's account of March 3–7 was written after returning to the ship.

19. Leffingwell Papers, Mss-69, Box 1, Folder 12.

20. Leffingwell Papers, Mss-69, Box 1, Folder 13.

21. Leffingwell did not provide details of the type of camera he carried, but his "glass lantern" slides are held at the U.S. Geological Survey in Denver.

22. Leffingwell Papers, Mss-69, Box 1, Folder 13.

23. RGS/LMS M 29 (Mikkelsen), Report of the Mikkelsen-Leffingwell expedition, 6. Royal Geographical Society (with Institute of British Geographers [IBG]).

24. RGS/LMS M 29 (Mikkelsen), Report of the Mikkelsen-Leffingwell expedition, 7. Royal Geographical Society (with IBG).

25. Leffingwell Papers, Mss-69, Box 1, Folder 13.

26. – 34. Ibid.

35. Mikkelsen, *Mirage in the Arctic*, 108.

36. Leffingwell Papers, Mss-69, Box 1, Folder 13.

Chapter 5

1. Leffingwell Papers, Mss-69, Box 1, Folder 13.

2. Ibid.

3. Ibid.

4. Charles Gil Mull is a geologist who worked for the U.S. Geological Survey and state agencies, and private industry, for many years in northern Alaska.

5. The author camped near there in 1991 and the sheep were still crossing the river at the same location.

6. Personal correspondence from Gil Mull to author. According to Mull: "Undoubtedly the reason they spent over a week at this location was the really interesting geology—a granite pluton—intrusion overlain by sedimentary rocks with some fossils," and known as the Okpilak Batholith. The Okpilak Batholith is composed

of igneous rock and extends across a significant portion of the Okpilak drainage; it was defined by E. G. Sable of the U.S. Geological Survey decades later.

7. Leffingwell Papers, Mss-69, Box 1, Folder 13.

8. – 10. Ibid.

11. According to Leffingwell's journal, Howe told Leffingwell that Storkerson had shot himself in the foot because he wanted to leave. In 1913, Storkerson told Johansen of the Canadian Arctic Expedition that Leffingwell had bribed him to leave.

12. Leffingwell Papers, Mss-69, Box 1, Folder 12.

13. – 16. Ibid.

17. Leffingwell Papers, Mss-69, Box 1, Folder 10.

18. Leffingwell Papers, Mss-69, Box 1, Folder 12.

19. Mikkelsen, *Conquering the Arctic Ice*, 337.

20. RGS/LMS M 29 (Mikkelsen), Summary of the Mikkelsen-Leffingwell expedition, Royal Geographical Society (with Institute of British Geographers [IBG]), 2.

21. Ibid., 4.

22. Ibid., 1.

23. Thomas Chrowder Chamberlin. Papers, Bound Letters, Vol. 21, no. 334, dated November 25, 1907, Special Collections Research Center, University of Chicago Library.

Chapter 6

1. Throughout his years on the North Slope, Leffingwell hired Inupiaq and non-Inupiaq to help him by hunting, cooking, cleaning, and transporting goods including provisions, equipment, and rocks, so that he could focus on his mapping. Some of the hired help assisted with mapping.

2. Leffingwell Papers, Mss-69, Box 1, Folder 13.

3. Leffingwell Papers, Mss-69, Box 1, Folder 12.

4. U.S. Geological Survey Geographic Names Information System, based on the National Elevation Dataset. See Introduction, Note 1.

5. Leffingwell Papers, Mss-69, Box 1, Folder 12.

6. – 10. Ibid.

11. Leffingwell Papers, Mss-69, Box 1, Folder 13.

12. Leffingwell respected the Inupiaq culture by naming many other geographic features based on his understanding of the Inupiaq pronunciation and meaning of the names. See Appendix 4.

13. Leffingwell Papers, Mss-69, Box 1, Folder 13.

14. Ibid.

15. Professional Paper 109, 135.

16. Leffingwell Papers, Mss-69, Box 1, Folder 12.

17. Leffingwell Papers, Mss-69, Box 1, Folder 13.

18. Leffingwell Papers, Mss-69, Box 1, Folder 12.

Chapter 7

1. Leffingwell Papers, Mss-69, Box 1, Folder 13.

2. – 7. Ibid.

8. Leffingwell Papers, Mss-69, Box 1, Folder 12.

9. Leffingwell's journal correctly stated July 17 as a Friday, but wrote over Saturday's date, not sure whether it was the 18th or 19th and identified Sunday as the 20th. His journal entries will be honored throughout even though the date is off one day. He corrected it August 17 in his journal entries.

10. Leffingwell Papers, Mss-69, Box 1, Folder 12.

11. The actual date was August 14.

12. Vilhjalmur Stefansson, *My Life with the Eskimo* (New York: Collier Books, 1971), 51.

13. Leffingwell Papers, Mss-69, Box 1, Folder 13.

14. Ibid.

15. Ibid.

Chapter 8

1. Special Collections and Archives, Knox College, Galesburg, Illinois. A yawl is a two-masted vessel with one mast located aft, and utilizes three or more sails.

2. Neill M. Clark, "Ships North to Alaska's Coast," in *Montana, The Magazine of Western History* 23, no. 4 (Autumn 1973), 37.

3. Leffingwell Papers, Mss-69, Box 2, Folder 1.

4. Ibid.

5. Leffingwell indicated miles for this entry in his journal, but did not indicate whether nautical or statute miles, and did use knots in other entries.

6. Leffingwell Papers, Mss-69, Box 2, Folder 1.

7. Ibid.

8. Ibid.

9. In 1913 Leffingwell would help pilot the *Mary Sachs* from Nome along the Alaskan coast to the North Slope for Stefansson's Canadian Arctic Expedition.

10. Leffingwell Papers, Mss-69, Box 2, Folder 1.

11. – 24. Ibid.

25. Gil Mull has indicated that the Flaxman Formation includes some exotic boulders carried by ice rafts from places as far away as Ellesmere Island or northern Greenland, and then deposited during a time of higher sea level. He explains that during

higher sea levels, the ice rafts grounded and left their deposits as the ice melted. Mull writes that the exotic boulders are "rock types not known to be present in the Brooks Range." Personal correspondence from Gil Mull to author.

26. Leffingwell Papers, Mss-69, Box 2, Folder 1.

27. Ibid.

28. Ibid.

Chapter 9

1. Leffingwell Papers, Mss-69, Box 2, Folder 2.

2. – 7. Ibid.

8. Personal correspondence from Gil Mull to author.

9. Leffingwell had considered calling the two lakes Weller and Williston, after Dr. Stuart Weller, geologist, and Williston, paleontologist, at the University of Chicago. According to Gil Mull, Peters and Schrader headed the first major USGS expedition across the Brooks Range in 1900-01.

10. Leffingwell Papers, Mss-69, Box 2, Folder 2.

11. Ibid.

12. Personal correspondence from Gil Mull to author. "The structural geology in most of the Sadlerochit Mountains area is relatively straightforward, but for several years we were also baffled by the geology at the mountain front near Marsh Creek. Finally we realized that the anomalous rocks are on thrust sheets that have been displaced northward a number of miles from the south side the mountains. This style of structural deformation is common in the western and central Brooks Range, but had not previously been recognized in the eastern Brooks Range; thus it is not surprising that Leffingwell was mystified."

13. Leffingwell Papers, Mss-69, Box 2, Folder 2.

14. – 19. Ibid.

Chapter 10

1. Leffingwell Papers, Mss-69, Box 2, Folder 3.

2. Ibid.

3. Leffingwell referred to it as Point Mikkelsen, but Bullen Point had been named by Sir John Franklin in 1828.

4. Leffingwell Papers, Mss-69, Box 2, Folder 3.

5. – 7. Ibid.

8. Leffingwell confused the dates in his journal because he labeled May 31 as June 1. He noted the discrepancy later in his journal. For the purposes of accuracy, the entries for June and July of 1912 will be identified under the correct date, which is one day later than he indicated in his original entries.

9. Leffingwell Papers, Mss-69, Box 2, Folder 3.

10. Ibid.

11. Ibid.

12. Leffingwell Papers, Mss-69, Box 2, Folder 15. *Burlington, Iowa Daily,* November 9, 1912. 4:00 p.m. edition.

Chapter 11

1. News article dated January 8, 1916, *Chicago Tribune*, Special Collections and Archives, Knox College.

2. Stefansson Papers, MSS-98 Stef Admin. Papers, Box 4, Folder 1, Dartmouth College Library, Rauner Special Collections.

3. "Incomplete report," p. 15, dated January 1, 1914, Stefansson Papers, MSS-98 Stef Admin. Papers Box 4, Folder 5, Dartmouth College Library, Rauner Special Collections.

4. Many accounts have been written about the roles of Stefansson, Captain Bartlett, the CAE, and the Karluk tragedy and are well worth reading.

5. Libraries and Archives Canada, Kenneth Gordon Chipman fonds, 1913-18, MG30, Diaries and Correspondence, series, B66, Vol 1. Diary #1 June 15, 1913 to March 5, 1914. August 17, 1913, entry.

6. Mary Sachs Island is now an extension of the Flaxman spit on the north side of Flaxman Island.

7. Libraries and Archives Canada, Kenneth Gordon Chipman fonds, 1913-18, MG30, Diaries and Correspondence, series, B66, CAE August–December 1913 Correspondence and Reports, letter from Kenneth Chipman to WH Boyd, Geological Survey Ottawa, September 7, 1913, p. 2.

8. Leffingwell Papers, Mss-69, Box 2, Folder 4.

9. Libraries and Archives Canada, John Johnston O'Neill fonds, MG30, Diaries and Correspondence, series, B171, Correspondence and Reports, letter from John O'Neill to Mr. Leroy, Geological Survey, Ottawa, October 13, 1913, p. 2.

10. Leffingwell Papers, Mss-69, Box 2, Folder 4.

11. Libraries and Archives Canada, Burt M. McConnell fonds, 1913-14 MG30, Diaries and Correspondence, series, B24, September 19, 1913 to April 7, 1914. December 11, 1913, entry.

12. Libraries and Archives Canada, Burt M. McConnell fonds, 1913–14, MG30, Diaries and Correspondence, series, B24, September 19, 1913, to April 7, 1914. December 12, 1913 entry.

13. Leffingwell Papers, Mss-69, Box 2, Folder 4.

14. Stefansson. *The Friendly Arctic*, 86.

15. Libraries and Archives Canada, Kenneth Gordon Chipman fonds, 1913–18, MG30, Diaries and Correspondence, series, B66, Vol 1. Diary #1 June 15, 1913 to March 5, 1914. December 25, 1913 entry.

16. Libraries and Archives Canada, Burt M. McConnell fonds, 1913–14, MG30, Diaries and Correspondence, series, B24, September 19, 1913 to April 7, 1914. February 3, 1914 entry.

17. Leffingwell Papers, Mss-69, Box 2, Folder 4.

18 – 20. Ibid.

21. Libraries and Archives Canada, Kenneth Gordon Chipman fonds, 1913–18, MG30, Diaries and Correspondence, series, B66, CAE January–June 1914 Correspondence and Reports, letter from Kenneth Chipman to Mr. LeRoy, April 30, 1914, p. 3.

22. Leffingwell Papers, Mss-69, Box 2, Folder 4.

23. Ibid.

24. Ibid.

25. Stefansson Papers, Mss-98, Series 13, Autobiography, 1906–1962, Box 58, F33.

Chapter 12

1. Executive Order No. 3797A, February 27, 1923. Establishing a Naval Petroleum Reserve in Alaska, Code of Federal Regulations.

2. U.S. Geological Survey, *Mineral Resources of Alaska*, 1908, 61–62.

3. Correspondence from Anna Leffingwell to Gil Mull, May 20, 1973. Gil Mull Files.

4. Most of the letters provide only the day of the week, and are not otherwise dated. Those with envelopes indicate the date mailed.

5. Deborah Storre Collection. All quotations from letters through page 234 are from this collection.

6. William Osler, a founder of Johns Hopkins Hospital, made reference in a humorous speech to Anthony Trollope's comments in the book, *The Fixed Period* (1882), which suggested euthanasia by chloroform after retirement.

7. "Done with the Wild and Icy North" and "Noted Arctic Explorer Will be Ranch Hand," Whittier, California, newspaper, n.d. Clipped articles in Leffingwell scrapbooks. Deborah Storre Collection.

8. The ranch was later sold and homes built in the early 1950s. An elementary school was built in 1955 and is named after Charles W. Leffingwell's son, Charles.

9. Correspondence from Anna Leffingwell to Gil Mull, June 9, 1973. Gil Mull Files.

10. Phone conversation and correspondence with Clive Rayne. September 12, 2009.

Chapter 13

1. Anna Leffingwell Diary, Deborah Storre Collection.

2. *Monterey Peninsula Herald*, October 2, 1957.

3. Leffingwell to Stefansson, October 5, 1962, Leffingwell Papers. Misc. Correspondence.

4. Evelyn Stefansson to Leffingwell, October 11, 1962, Leffingwell Papers, Misc. Correspondence.

5. Deborah Storre Collection.

6. *Monterey Peninsula Herald*, December 21, 1967, 5.

7. The Explorers Club of New York to Anna Leffingwell, April 5, 1971. Deborah Storre Collection.

8. Anna Leffingwell to Gil Mull, April 9, 1971. Gil Mull Files.

9. Deborah Storre Collection.

10. John F. [Jack] Wilson February 19, 1971, correspondence. Deborah Storre Collection.

11. Deborah Storre Collection.

12. Personal correspondence from Terry Grebens to author, n.d.

13. *The Pine Cone*, Carmel by the Sea, CA, January 4, 1973, 6, 7, 14

14. John M. Kauffmann, *Alaska's Brooks Range* (Seattle: Mountaineers Books, 1992), 63.

Chapter 14

1. *Chicago Tribune*, January 8, 1916, Special Collections and Archives, Knox College.

2. Correspondence from Royal Geographical Society to Leffingwell, May 3, 1922, Deborah Storre Collection. In August 1923, the Royal Geographical Society requested a photo from Leffingwell for hanging in the museum. Leffingwell submitted a photo that had been taken just after he had returned from Alaska in 1914. The photograph is mounted on the wall with other recipients at the Royal Geographical Society.

3. Leffingwell Papers, Mss-69, Box 2, Folder 15, *Los Angeles Times*, July 16, 1922.

4. Leffingwell Papers, Mss-69, Box 1, Folder 8.

5. *Knox Alumnus*, April–May 1923. Special Collections and Archives, Knox College.

6. *Pasadena Star News*, March 31, 1923. Deborah Storre Collection.

7. Correspondence from Secretary, The Explorers Club, to Leffingwell, November 20, 1922. The Explorers Club Archives, The Explorers Club, New York.

8. Correspondence from Hillard N. Reiser, U.S. Geological Survey, to Mrs. Ernest de K. Leffingwell. December 5, 1972. Deborah Storre Collection.

9. Personal correspondence from Gil Mull to Ernest Leffingwell. Gil Mull Files.

10. Plaque at Flaxman Island site. Gil Mull Files.

11. Personal correspondence from Gil Mull to author.

Epilogue

1. Professional Paper 109, 179.

2. Professional Paper 109, 37.

3. Personal correspondence from Kenji Yoshikawa to author, September 1, 2014.

Bibliography

Banerjee, Subhankar, ed. *Arctic Voices: Resistance at the Tipping Point.* New York: Seven Stories Press, 2012.

Banerjee, Subhankar, and Peter Matthiessen. *Arctic National Wildlife Refuge: Seasons of Life and Land: A Photographic Journey.* Seattle: Mountaineers Books, 2003.

Berton, Pierre. *Prisoners of the North: Portraits of Five Arctic Immortals.* New York: Carroll & Graf Publishers, 2004.

Brinkley, Douglas. *The Quiet World: Saving Alaska's Wilderness Kingdom 1879–1960.* New York: Harper Collins, 2011.

Brooks, Alfred H. *Blazing Alaska's Trails.* College: University of Alaska and the Arctic Institute of North America, 1953.

Brower, Charles D., in collaboration with Philip J. Farrelly and Lyman Ansom. *Fifty Years Below Zero: A Lifetime of Adventure in the Far North.* New York: Dodd Mead & Co., 1943.

Brown, Stephen C. *Arctic Wings: Birds of the Arctic National Wildlife Refuge.* Seattle: Mountaineers Books, 2006.

Capelotti, P. J. "A 'Radically New Method': Balloon Buoy Communications of the Baldwin-Ziegler Polar Expedition, Franz Josef Land, June 1902." *Polar Research* 27 (February 2008): 52–72.

_____. *The Greatest Show in the Arctic: The American Exploration of Franz Josef Land, 1898–1905.* Norman: University of Oklahoma Press, 2016.

Caswell, John Edwards. *Arctic Frontiers: United States Explorations in the Far North.* Norman: University of Oklahoma Press, 1956.

Clark, Neil M. "Ships North to Alaska's Coast." *Montana, The Magazine of Western History* 23, no. 4 (Autumn 1973): 32–41.

Collier, Michael. *Sculpted by Ice: Glaciers and the Alaska Landscape.* Anchorage: Alaska Natural History Association, 2004.

Craighead, Charles, and Bonnie Kreps. *Arctic Dance: The Mardy Murie Story.* Portland, OR: Graphic Arts Center Publishing, 2002.

Delgado, James P. *Across the Top of the World: The Quest for the Northwest Passage.* New York: Checkmark Books, 1999.

Fogelson, Nancy. *Arctic Exploration & International Relations, 1900–1932: A Period of Expanding National Interests.* Fairbanks: University of Alaska Press, 1992.

Franklin, Sir John. *Sir John Franklin's Journals and Correspondence: The Second Arctic Land Expedition, 1825–1827.* Edited and with an introduction by Richard Clarke Davis. Toronto: Champlain Society, 1998.

Freed, Stanley A. "Fate of the Crocker Land Expedition." *Natural History* (June 2012): 10–19.

Fritts, Crawford E. and Mildred E. Brown. *Bibliography of Alaskan Geology, Vol. 1 (1831–1918)*. College: Alaska Division of Geological Survey, 1971.

Hanable, William S. "Leffingwell: Prudhoe's Pioneer Scientist." *Exxon USA* 12, no. 1 (1973): 28, 30.

Harrison, Alfred H. *In Search of a Polar Continent, 1905–1907*. London: E. Arnold, 1908.

Hayes, Derek. *Historical Atlas of the Arctic*. Seattle: University of Washington Press, 2003.

Henighan, Tom. *Vilhjalmur Stefansson: Arctic Adventurer*. Toronto: Dundurn Press, 2009.

Henderson, Joe. "Retracing Leffingwell's Steps." *Mushing Magazine*. Part 1, July/Aug 2006; Part 2, Sept. 2007; Part 3, Nov. 2008.

Holland, Clive. *Arctic Exploration and Development, c. 500 B.C. to 1915: An Encyclopedia*. New York: Garland Publishing, 1994.

International Boundary Commission (U.S., Alaska, and Canada). *Joint Report Upon the Survey and Demarcation of the International Boundary Between the United States and Canada Along the 141st Meridian from the Arctic Ocean to Mount St. Elias*. Signed at Washington, April 21, 1906.

Jayne, A. G. *Frozen Justice: A Story of Alaska*. Translated from the Danish of Ejnar Mikkelsen. New York: Alfred A. Knopf, 1922.

Jenness, Stuart E. ed. *Arctic Odyssey: The Diary of Diamond Jenness, Ethnologist with the Canadian Arctic Expedition in Northern Alaska and Canada, 1913–1916*. Hull, Quebec: Canadian Museum of Civilization, 1991.

Jones, Jeff. *Arctic Sanctuary: Images of the Arctic National Wildlife Refuge*. Fairbanks: University of Alaska Press, 2010.

Kauffmann, John M. *Alaska's Brooks Range: The Ultimate Mountains*. Seattle: Mountaineers Books, 1992.

Kaye, Roger. *Last Great Wilderness: The Campaign to Establish the Arctic National Wildlife Refuge*. Fairbanks, AK: University of Alaska Press, 2006.

Lainema, Matti, and Julia Nurminen. *A History of Arctic Exploration: Discovery, Adventure and Endurance at the Top of the World*. London: Conway; NY: Sterling, 2009.

Leffingwell, Ernest deKoven. "The Anglo-American Polar Expedition." *National Geographic Magazine*, December 1907, 796.

_____. "A Brief Account of the Baldwin-Ziegler Polar Expedition." *The File Closer*. June 1903.

_____. "Camping Alone by the Frozen Sea." *Colliers*, March 13, 1909, 19–20.

_____. *The Canning River Region, Northern Alaska*. U.S. Geological Survey Professional Paper, 109. Washington, DC: Government Printing Office, 1919. pubs.usgs.gov/pp/0109/report.pdf.

_____. "A Communication from Leffingwell," *The University of Chicago Magazine*. January 1915, 76–79.

_____. "Flaxman Island, A Glacial Remnant." *Journal of Geology* 16 (January 1908): 56–63.

_____. "Ground-ice Wedges: The Dominant Form of Ground Ice on the North Coast of Alaska." *Journal of Geology* 23, no. 7 (October–November 1915): 635–54.

_____. "My Polar Explorations 1901–1914." *Explorers Journal* 39, no. 3 (October 1961): 2–14.

_____. "Pleistocene Geology of the Sawatch Range Near Leadville, Colo." *Journal of Geology*. 12, no. 8 (1904): 698–706.

_____. "A Reconnaissance of the Arctic Slope of Alaska" (Abstract). *Washington Academy of Science Journal* 3, no. 11 (June 4, 1913): 343–44. Paper presented at the 269th meeting of The Geological Society of Washington, April 9, 1913.

Lopez, Barry Holstun. *Arctic Dreams: Imagination and Desire in a Northern Landscape*. New York: Bantam Books, 1989.

Marshall, Robert. *Arctic Wilderness*. Berkeley: University of California Press, 1956.

McCannon, John. *A History of the Arctic: Nature, Exploration and Exploitation*. London: Reaktion Books Ltd., 2012.

McPhee, John. *Coming into the Country*. New York: Farrar, Straus and Giroux, 1977.

Mikkelsen, Ejnar. *Mirage in the Arctic: Explorations in Unknown Alaska*. Introduction by Lawrence Millman. Lyons Press Arctic Adventure Classic. Guilford, CT: Lyons Press, 2005.

_____, with Ernest de K. Leffingwell and G. P. Howe. *Conquering the Arctic Ice*. London: W. Heinemann, 1909.

Miller, Debbie S. *Midnight Wilderness: Journeys in Alaska's Arctic National Wildlife Refuge*. San Francisco: Sierra Club Books, 1990.

Milton, John P. *Nameless Valleys, Shining Mountains: The Record of an Expedition into the Vanishing Wilderness of Alaska's Brooks Range*. New York: Walker and Co., 1970.

Mountfield, David. *A History of Polar Exploration*. New York: The Dial Press, 1974.

Mull, Charles G. "Leffingwell Pioneer Arctic Explorer-Scientist." *Alaska Geographic: The North Slope* 1, no. 1, (1972): 56-58.

Mull, Gil. "Mystic Mountains." *Alaska Geographic* 23, no. 3 (1996): 36.

_____. "Mystic Mountains with a Rich History." *Alaska Geographic: The Brooks Range Environmental Watershed* 4, no. 2 (1977): 13–33.

Murie, Margaret E. *Two in the Far North*. New York: A. A. Knopf, 1967.

National Geographic Magazine. *The Arctic Number* 18, no. 7 (July 1907). Washington, DC: The National Geographic Society, 1907.

Niven, Jennifer. *The Ice Master: The Doomed 1913 Voyage of the Karluk*. New York: Hyperion, 2000.

North Slope Borough Commission on History & Culture. *Dogsled Trip from Barrow to Demarcation Point, April 1937: Diary of Fred G. Klerekoper*. June 1977.

Operti, Albert, compiler. "A Chronological List of Arctic and Antarctic to the Poles North and West Passages and Relief Expeditions." *Arctic Club of America Collection 1906*. The Explorers Club Archives. Box 1 of 4. Folder Title: Manual 1906. New York: The Arctic Club, 1906.

Orth, Donald J. *Dictionary of Alaska Place Names*. U.S. Geological Survey Professional Paper 567, Washington, DC: United States Government Printing Office, 1967.

Pielou, E. C. *A Naturalist's Guide to the Arctic*. Chicago: University of Chicago Press, 1994.

Pratt, Kenneth L., ed. *Chasing the Dark: Perspectives on Place, History and Alaska Native Land Claims*. Shadowlands, Vol. 1. Anchorage: U.S. Bureau of Indian Affairs, Alaska Division, 2009.

Reed, J. C., 1958, *Exploration of Naval Petroleum Reserve No. 4 and Adjacent Areas, Northern Alaska, 1944–1953*. Part 1. History of the Exploration. U.S. Geological Survey Professional Paper 301. Washington, DC: Government Printing Office, 1958.

Reid, Robert Leonard. *Arctic Circle: Birth and Rebirth in the Land of the Caribou*. Boston: David R. Godine, 2010.

Rennick, Penny, ed. "Opening the Arctic" and "Geology of the Arctic Slope." *Alaska Geographic: North Slope Now* 16, no. 2 (1989): 42–68.

Robinson, Michael F. *The Coldest Crucible: Arctic Exploration and American Culture*. Chicago: University of Chicago Press, 2006.

Roderick, Jack. *Crude Dreams*. Fairbanks, AK: Epicenter Press, 1997.

Sable, E. G. *Geology of the Romanzof Mountains, Brooks Range, Northeastern Alaska*. U.S. Geological Survey Professional Paper 897. Washington, DC: Government Printing Office, 1965. pubs.usgs.gov/of/1983/0578/report.pdf.

Sale, Richard. *Polar Reaches: The History of Arctic and Antarctic Exploration*. Seattle: The Mountaineers Books, 2002.

Sale, Richard, and Eugene Potapov. *The Scramble for the Arctic: Ownership, Exploitation and Conflict in the Far North*. London: Frances Lincoln, 2010.

Schrader, F. C., and W. J. Peters. *A Reconnaissance in Northern Alaska*. U.S. Geological Survey Professional Paper: 20. Washington, DC: Government Printing Office, 1904. pubs.usgs.gov/pp/0020/report.pdf.

Smith, Richard G. "Alaskan Oil and National Wildlife Ranges." *Association of Pacific Coast Geographers Yearbook* 35 (1973): 75–85.

Stefansson, Vilhjalmur. *Discovery: The Autobiography of Vilhjalmur Stefansson*. New York: McGraw-Hill Book Company, 1964.

______. *The Friendly Arctic: The Story of Five Years in Polar Regions*. New York: The Macmillan Co., 1921.

______. *My Life with the Eskimo*. New York: Collier Books, 1913. First Collier Books Ed., 1962. Third printing, 1971.

______. *Polar Expedition Diaries of Vilhjalmur Stefansson in the Years 1906–1918*. Ann Arbor, MI: Xerox University Microfilms; Hanover, NH: Dartmouth College Library, 1974.

______. *Report of the Canadian Arctic Expedition, 1913–18*. Ottawa: F.A. Acland, 1919–1946.

______. *Vilhjalmur Stefansson Diaries: Anglo-American Polar Expedition, 1906–1908, and also Stefansson-Anderson Expedition, 1908–1912*. Ann Arbor, MI: Xerox University Microfilms in collaboration with Dartmouth College Library, 1974.

Stuck, Hudson. *A Winter Circuit of Our Arctic Coast: A Narrative of a Journey with Dog-sleds around the Entire Arctic Coast of Alaska*. New York: C. Scribner's Sons, 1920.

Sweet, John M. *Discovery at Prudhoe Bay: Oil: Mountain Men and Seismic Vision Drilled Black Gold*. Blaine, WA: Hancock House Publishers, 2008.

Taliaferro, John. *In a Far Country: The True Story of a Mission, a Marriage, a Murder, and the Remarkable Reindeer Rescue of 1898*. New York: Public Affairs, 2006.

Todd, Flip. "Flaxman Island Back in 1907…" *Alaska Industry* (October 1978): 33–50.

Ulibarri, George S. *Documenting Alaskan History: Guide to Federal Archives Relating to Alaska*. Alaska Historical Commission Studies in History, no. 23. Fairbanks: University of Alaska Press, 1982.

United States Geological Survey. *Mineral Resources of Alaska: Report on Progress of Investigations in 1908*. Bulletin 379. Washington, DC: Government Printing Office, 1909, 61–62.

_____. *Summary Report for 1915*. Washington, DC: Government Printing Office, 1916, 222.

U.S. Fish and Wildlife Service. *Arctic National Wildlife Refuge, Alaska, Coastal Plain Resource Assessment: Report and Recommendation to the Congress of the United States and Final Legislative Environmental Impact Statement*. Washington, DC: U.S. Department of the Interior, 1987. pubs.usgs.gov/fedgov/70039559/report.pdf.

U.S. Fish and Wildlife Service. Region 7. *Arctic National Wildlife Refuge: Final Comprehensive Conservation Plan, Environmental Impact Statement, Wilderness Review, and Wild River Plans*. Anchorage: U.S. Fish and Wildlife Service, Region 7, 1988.

Waterman, Jonathan. *Arctic Crossing: A Journey Through the Northwest Passage and Inuit Culture*. New York: Knopf, 2001.

_____. *Where Mountains are Nameless: Passion and Politics in the Arctic National Wildlife Refuge: Including the Story of Olaus and Mardy Murie*. New York: Norton, 2005.

Watkins, T. H. *Vanishing Arctic: Alaska's National Wildlife Refuge*. New York: Aperture, 1988.

Wickersham, James. *A Bibliography of Alaskan Literature, 1724–1924*. Misc. Pubs. Vol. 1. Cordova: Alaska Agricultural College and School of Mines, 1927.

Wilkinson, Alec. *The Ice Balloon: S. A. Andree and the Heroic Age of Arctic Exploration*. New York: Alfred A. Knopf, 2011.

Williams, Glyndwr. *Arctic Labyrinth: The Quest for the Northwest Passage*. Berkeley: The University of California Press, 2010.

Wright, John Kirtland. *Geography in the Making: The American Geographical Society 1851–1951*. New York: American Geographical Society, 1952.

Young, Steven B. *To the Arctic: An Introduction to the Far Northern World*. New York: John Wiley & Sons, Inc., 1987.

Selected List of Maps Consulted

Arctic National Wildlife Refuge [computer map]. 1:600,000. AK: Borealis Maps, date not given.

Canada Natural Resources. *North Circumpolar Region* [map]. 2008. 1:9,000,000. The Atlas of Canada; MCR0001. Ottawa, Ontario: Natural Resources Canada, 2008.

Geological Society of America. *Geologic Sections and Maps across Brooks Range and Arctic Slope to Beaufort Sea, Alaska* [map]. 1987. 1:500,000. Map and Chart Series; MC-28S. Boulder, CO: The Geological Society of America, 1987.

International Travel Maps. *Alaska*. Third Edition. 1:2,500,000. Vancouver, BC: ITMB, 2000.

_____. *Nunavut*. First Edition. 1:1,850,000. Vancouver, BC: ITMB, 2009.

Jackson, Frederick G. *Western Franz Josef Land: Up to September 1895* [map]. 1:750,000. Map No. 2. New York & London: Harper & Brothers, Publishers.

_____. *Sketch Map of Franz Josef Land Showing Journeys & Discoveries of Frederick Jackson F.R.G.S. Leader of the Jackson-Harmsworth Polar Expedition* [map]. Scale not given. London: George Philip & Sons, 1895.

_____. *Map of Franz Josef Land Showing Discoveries of Frederick G. Jackson F.R.G.S., Commander of the Jackson-Harmsworth Polar Expedition, 1894–7* [map]. 1,000.000. Compiled from: Copeland-Payer's, Leigh-Smith's, Nansen's, and Jackson's Maps. N.p.: nd.

Nansen, Fridtjof. *Map Showing the Route of the "Fram" and Nansen's and Johansen's Sledge Journey* [map]. Scale not given. In Nansen's *Farthest North*. New York: Harper & Brothers, 1897.

_____. *Preliminary Sketch Map of the Group of Islands Known as Franz Josef Land* [map]. Scale not given. "Compiled at Cape Flora, July 1896, and based upon Payer's, Leigh Smith's, and Jackson's Maps, together with my own observations." In Nansen's *Farthest North*. New York: Harper & Brothers, 1897.

National Geographic Magazine. "Map of the North Pole Regions" [map]. Supplement to the *National Geographic Magazine*, July 1907. Scale [ca. 1:13,500,000]. Washington, DC: The National Geographic Society, 1907.

Payer, Julius. *Endgultige Karte Von Franz Josef Land. 2. Osterr.-Ungar. Nordpolar-Expedition 1873 & 1874* [map]. 1:1,000,000. In Petermann's *Geographische Mittheilungen*. Tafel 11. Gotha: Justus Perthes, 1876.

_____. *Originalkarte Des Kaiser Franz Josef Landes* [map]. Scale not given. Map 111. N.p., 1874.

_____. *Zweite Provisorische Karte Von Franz Josef Land: Osterr.-Ungar. Nordpolar-Expedition 1873 & 1874* [map]. 1:1,600,000. In Petermann's *Geographische Mittheilungen* 10. Okt. 1874, Jahrgang 1874, Tafel 23. Gotha: Justus Perthes, 1874.

Porter, R. W. *Map of Franz Josef Archipelago* [map]. 1:750,000. Compiled from surveys of the Ziegler Polar Expeditions 1901–02, 1903–05, Map B, 1907.

Smith, B. Leigh *Discoveries along the South Coast of Franz-Josef Land* [map]. Scale not given. 1880. Published for the Proceedings of the Royal Geographical Society, 1881.

U.S. Army Map Service *Tromso, Norway* [map]. 1:50,000. Sheet 1534-3. AMS Series M711, (GSGS 4246). First Edition. Transverse Mercator Projection. Washington, DC: Army Map Service, 1950 (reprinted: Norges Geografiske Oppmaling, 1967).

U.S. Defense Mapping Agency. *Tactical Pilotage Chart, TPC A-2D* [map]. Edition 1, 1988. 1:500,000. Polar Stereographic Projection. St. Louis, MO: Defense Mapping Agency Aerospace Center, 1988.

U.S. Geological Survey. *Canning River Region Northern Alaska*. Professional Paper 109. Washington, DC: Government Printing Office, 1919. (Maps by Ernest Leffingwell.)

_____. Plate 1. Reconnaissance map of the Canning River region, Alaska.

_____. Plate 2. Geologic reconnaissance map of Canning River region, Alaska.

_____. Plate 3. Map of the north Arctic coast, Alaska.

_____. Plate 4. Map of the coast line between Challenge Entrance and Thetis Island, Alaska.

_____. Plate 5. Map of the coastline between Martin Point and Challenge Entrance, Alaska.

Plate 10. Triangulation stations along the Arctic coast adjacent to Canning River, Alaska.

_____. *Arctic, Alaska* [map]. 1956, Limited Revision, 1983. 1:250,000. Topographic Series. Reston, VA: United States Department of the Interior, USGS, 1988.

_____. *Barter Island (A-5) quadrangle, Alaska* [map]. 1955, Minor Revisions, 1985. 1:63,360. Topographic Series. Reston, VA: United States Department of the Interior, USGS, 1985.

_____. *Demarcation Point, Alaska* [map]. 1955, Limited Revision, 1983. 1:250,000. Topographic Series. Reston, VA: United States Department of the Interior, USGS, 1983.

_____. *Demarcation Point (B-5) quadrangle, Alaska* [map]. 1956, 1:63,360. Topographic Series. Reston, VA: United States Department of the Interior, USGS, 1966.

_____. *Flaxman Island, Alaska* [map]. 1955, Limited Revision, 1983. 1:250,000. Topographic Series. Reston, VA: United States Department of the Interior, USGS, 1989.

_____. *Flaxman Island (A-4) quadrangle, Alaska* [map]. 1955, Limited Revision, 1981. 1:63,360. Topographic Series. Reston, VA: United States Department of the Interior, USGS, 1981.

_____. *Geologic Map of the Demarcation Point, Mt. Michelson, Flaxman Island, and Barter Island Quadrangles, Northeastern Alaska* [map]. 1986. 1:250,000. Geologic Investigations Series; I-1791. Reston, VA: United States Department of the Interior, USGS, 1986.

_____. *Mt. Michelson, Alaska* [map]. 1956, Limited Revision, 1983. 1:250,000. Topographic Series. Reston, VA: United States Department of the Interior, USGS, 1983.

_____. *Mt. Michelson (B-1) quadrangle, Alaska* [map]. 1955, 1:63,360. Topographic Series. Reston, VA: United States Department of the Interior, USGS, 1966.

_____. *Mt. Michelson (C-2) quadrangle, Alaska* [map]. 1955, 1:63,360. Topographic Series. Reston, VA: United States Department of the Interior, USGS, 1975.

_____. *State of Alaska (Map B, East Half)* [map]. 1973, Revised, 1986. 1:1,584,000. Reston, VA: United States Department of the Interior, USGS, 1987.

_____. *State of Alaska (Map E)* [map]. 1973, Revised, 1996. 1:2,500,000. Reston, VA: United States Department of the Interior, USGS, 1996.

U.S. National Imagery and Mapping Agency. *Jet Navigation Chart, JNC 4* [map]. Edition 3, 1998. 1:2,000,000. Transverse Mercator Projection. Bethesda, MD: National Imagery and Mapping Agency, 1998.

U.S. National Oceanic and Atmospheric Administration. National Ocean Service [chart]. 1: *Alaska Nautical Charts*. Rockville, MD: National Ocean Service, dates and editions vary. Alaska Charts: 16004, 16041, 16042, 16043, 16044, 16045, 16046, 16061, 16062, 16063, 16064, 16065, 16066, 16067, 16081, 16082.

Weyprecht, Karl. *Originalkarte Der Ruckreise Der Osterr.-Ungar. Expedition Mai–August 1874* [map]. 1:1,900,000. Nach's Weyprecht Beobachtungen von A. Petermann. In *Petermann's Geographische Mittheilungen*. Jahrgang 1877, Tafel 5. Gotha: Justus Perthes, 1877.

Archives Consulted

American Geographical Society of New York Archives, American Geographical Society Library, University of Wisconsin-Milwaukee Libraries (AGS Archives)

Danish Arctic Institute, Copenhagen, Denmark

Danish Royal National Library, Copenhagen, Denmark

Dartmouth College, Rauner Library Special Collections, Stefansson Collection, Hanover, NH

Deborah Storre Private Collection, Eureka, CA

The Explorers Club Archives, New York, NY

Knox College, Special Collections and Archives, Galesburg, IL

Knoxville Public Library, Knoxville, IL

Libraries and Archives Canada, Ottawa, Ontario

Library of Congress, Manuscript Division, Library of Congress, Washington, DC

Monterey Public Library, Monterey, CA

National Museum of History, Canada [Canadian Museum of History, Gatineau, Quebec]

National Park Service, United States

Pioneers Museum and Cultural Center of the Imperial Valley (formerly Imperial Valley Historical Society), Imperial, CA

Royal Geographical Society, London

Smithsonian Archives, Washington, DC

Seattle Public Library, Seattle, WA

Terry Grebens Private Collection, Nevada City, CA

Trinity College, Watkinson Library, Hartford, CT

University of Alaska, Alaska & Polar Regions Collections and Archives, Rasmuson Library, Fairbanks, AK

University of Chicago Library, Special Collections Research Center, Chicago, IL

University of Victoria Libraries, Victoria, British Columbia

University of Washington, Special Collections, University Library, Seattle, WA

U.S. Geological Survey, Anchorage, AK, Leffingwell Scientific Notebooks

U.S. Geological Survey, Denver, CO, Leffingwell Digital Images and Photocopies

U.S. Geological Survey, Fairbanks, AK, Flaxman Island Aerial Photographs

Victoria City Archives, Victoria, British Columbia

Interviews

Terry Grebens, Nevada City, California, personal communication with author (undated).

Gil Mull, Santa Fe, New Mexico, personal correspondence with author (undated).

Clive Rayne, Carmel, California, conversation with author (undated), correspondence, September 12, 2009.

Deborah Storre, Eureka, California, personal communication with author (undated).

Websites Consulted

Alaska Department of Natural Resources, dnr.alaska.gov

Alaska Department of Natural Resources, Alaska Division of Geological & Geophysical Surveys, www.dggs.alaska.gov/pubs/id/3803

American Geographical Society, americangeo.org

Arctic Institute of North America, arctic.ucalgary.ca

Arctic National Wildlife Refuge, www.fws.gov/refuge/arctic

Canadian Museum of History, www.historymuseum.ca

Canadian Geographical Names, www.geobase.ca

Dartmouth College. Rauner Library. Leffingwell Collection, ead.dartmouth.edu/html/stem69.html

The Explorers Club, www.explorers.org/index.php/about/history/a_gathering_place

Latitude/Longitude Distance Calculator, www.nhc.noaa.gov/gccalc.shtml

Library and Archives Canada, www.bac-lac.gc.ca/eng/Pages/home.aspx

National Geospatial Agency, GEOnet Names Server (GNS), geonames.nga.mil/gns/html

National Oceanic and Atmospheric Administration, National Ocean Service, oceanservice.noaa.gov

National Oceanic and Atmospheric Administration, National Weather Service, www.weather.gov/ama/conversions#tables

National Oceanic and Atmospheric Administration, Office of Coast Survey, www.charts.noaa.gov/InteractiveCatalog/nrnc.shtml

Royal Geographic Society, www.rgs.org/AboutUs/Medals+and+awards

Scott Polar Research Institute, www.spri.cam.ac.uk

Smithsonian National Museum of Natural History, collections.mnh.si.edu/search

Stefansson Collection, www.dartmouth.edu/~library/rauner/manuscripts/stefansson_guide.html

U.S. Board on Geographic Names, GNIS, geonames.usgs.gov

U.S. Fish and Wildlife Service, www.fws.gov

U.S. Geological Survey, www.usgs.gov

U.S. Geological Survey National Geologic Map Database, ngmdb.usgs.gov/Geolex

U.S. Geological Survey Denver Library Photographic Collection, https://library.usgs.gov/photo/#

U.S. National Park Service, www.nps.gov

Index

Page numbers in italics refer to illustrations.

About the Author

Photo by Dan Bowles

Janet R. Collins has visited Alaska and the North Slope numerous times, backpacking and rafting rivers in the Arctic National Wildlife Refuge and in the Gates of the Arctic National Park since 1991. Her passion for the arctic environment and its history date back to an arctic environment college course she took in the 1970s. She has an undergraduate degree in geography, a master's degree in library science, and is a graduate of the Eden Energy Medicine Certification Program. She was director and map librarian at the Huxley Map Library at Western Washington University.